THE DOW JONES-IRWIN GUIDE TO

HIGH TECH INVESTING

PICKING TOMORROW'S WINNERS TODAY

THE DOW JONES-IRWIN GUIDE TO

HIGH TECH INVESTING

PICKING TOMORROW'S WINNERS TODAY

JAMES B. POWELL

DOW JONES-IRWIN Homewood, Illinois 60430

ISBN 0-87094-596-3

Library of Congress Catalog Card No. 85–72178

Printed in the United States of America

1 2 3 4 5 6 7 8 9 0 K 3 2 1 0 9 8 7 6

*To my family,
my greatest
treasure.*

PREFACE

ADDRESSING A NEED: A BOOK OF MAJOR TECHNOLOGICAL TRENDS AND OPPORTUNITIES FOR INVESTORS

We are living in a period of rapid technological change that affects millions of people, often in very significant ways. Not surprisingly, many excellent books have been written to help us understand and cope with this change. The best of these books can make it possible for many individuals and organizations not only to survive, but also to prosper.

I've found from talking to many investors, however, that although the popular books about the future leave them better informed, they also leave them frustrated. While the books may identify major social and technological changes, they don't examine the changes from an investor's viewpoint. They provide valuable insights and background information, but lack specific money-making suggestions. One client summed up the popular futurist books by saying, "For an investor, reading these books is like walking down a street catching whiffs of magnificent foods being cooked that you can only partly identify and can't locate."

Following Through

This book is different. It is about major technological changes and it is written specifically for investors. The book identifies and explains the significance of each trend, then follows up with pertinent investment advice. Each chapter concludes with a review of promising public companies that are most likely to profit from the changes previously discussed.

Many Considerations, Few Selections

This book is very selective. I've only presented areas in which investors can take profitable positions that will begin to show returns within a reasonable length of time. Other changes that will have a large impact on our lives but are not presently available in the market are similarly omitted.

Just as my selection criteria require the rejection of exciting areas that have no investment potential, they also require me to present subjects that are very promising even though they are temporarily out of favor with some investors. Major technological developments such as robotics are frequently written up years before they have any chance of showing a profit and invariably disappoint scores of investors. Later, when the technology quietly matures and begins to fulfill its promise, no one pays any attention. Several top-notch turnaround technologies are presented on these pages that will outperform others with more "sizzle."

The book also contains several sections about specific types of investments that will make it easier to create portfolios suited to your needs. I include chapters on selecting high tech growth stocks, mutual funds, and new issues. In addition, I discuss venture capital investments and technology index options.

The book concludes with a unique directory of over 1,250 U.S. and foreign public high tech companies. Each listing contains the firm's stock symbol, its exchange, and appropriate contact information. Investors should find the directory very useful in following up companies mentioned in this book and in the news.

The major technological trends discussed in the following pages will return billions of dollars to knowledgeable investors over the next few years. Readers who take the time to become acquainted with the trends and the specific opportunities they present should be well rewarded for their efforts.

ACKNOWLEDGMENTS

Alexander P. Paris, well known free market economist, money manager and president of Barrington Research Associates, Barrington, Illinois has provided me with many valuable insights regarding market strategy and economic timing which have become an important part of this book. His contribution helped me to make this volume a specific guide to successful investing instead of a more general book about trends in technology.

I wish also to thank James U. Blanchard III and David W. Galland for providing me with the first public forum to present my investment analysis and strategies. My involvements with the annual NCMR conventions, *Wealth* Magazine, and the Wealth Institute have been most rewarding.

A very special acknowledgment is due to my friend and associate, Robert H. Meier. His advice and encouragement over the years have substantially enriched my life.

James B. Powell

ABOUT THE AUTHOR

James B. Powell has an extensive background in both the sciences and finance which has made his analysis of high tech trends and companies highly valued among professional investors. Mr. Powell is Managing Editor of two newsletters, *High Tech Investor* and *Technology Stock Monitor*, which are published by Barrington Research Associates, P.O. Box 860, Barrington, Illinois 60010.

Mr. Powell is also President of Technology Investment Research, Inc., P.O. Box 526, DeKalb, Illinois 60115, a company that specializes in consulting services to the investment community, technology companies, and venture capital firms.

CONTENTS

PART II
Investment Vehicles and Strategies

Section 11
Choosing the Type of Investment That's Best for You

INTRODUCTION

The Case for High Tech Investing

If you look at the history of investments in the United States since the Industrial Revolution, you will quickly notice that real estate and technology have consistently led the pack. The margin by which technology leads other investments appears to be getting even wider. Real estate began to level out several years ago and is not expected to begin returning significant profits again before the 1990s.

Neither is the technology movement likely to fade away. Both as individuals and as a society we are so highly dependent on advanced technology that we could not survive without it. As the world's population continues to expand while its resources diminish, our dependence on technology will become even greater.

Sophisticated technology is also the hope of the Third World countries, where the lack of development is erroneously seen by some people as a blessing in disguise. Third World countries need high tech agriculture and industrial production desperately if they are to survive with even a minimal level of human dignity. "We can go back to a simpler age," said a former secretary of agriculture, "but first we need to decide which half of the population will be allowed to die."

Before getting into other matters, I should point out that investing in the great majority of high technology companies has noneconomic rewards as well as the potential for high profits. Although I take a cold-hearted view toward every investment and do not allow anything but performance and good research to influence my decisions, I am also aware that I am investing in ideas and people that solve important problems. When I move money from underperforming companies to those with greater promise, I know I am placing my wealth where it will do the most good. To be sure, my motive for making any investment is to increase my profits, but it is nevertheless satisfying to do so in fields that are improving the quality of life for millions of people.

KNOWLEDGE IS THE KEY TO SUCCESS

Every investor makes the occasional lucky shot. Every once in awhile you put your money on a company that you don't have time to investigate very well and the company performs spectacularly. Unfortunately, lucky shots can't be sustained and great wealth rarely comes from them.

A much surer path to success can be found by compounding the profits from a series of smaller gains made over several years. Unlike the occasional lucky shot, however, consistent winning takes a bit of study and attention. Fortunately, with technology investments the rewards balance the efforts and the information itself is fascinating.

At the present time, informed high tech investors are an exclusive minority with a distinct advantage in the marketplace. Few investors have the foresight you are demonstrating and are babes in the woods when it comes to making good decisions about technology investments. I can't think of any other investment area where a little homework can place you so far ahead of the competition. I'd congratulate you for your commitment but I would rather let the market do that job for me. Your awards and honors will take the form of successful trades and compounded profits.

YOU DON'T NEED TO BE AN ENGINEER

You don't need a technical background to become a successful high tech investor. In fact, a heavy technological background can be a liability, because people who have one frequently assume that top-notch products automatically mean top-notch profits. Technical people too often ignore the business and psychological aspects of companies and markets and lose money as a result. People who are fixated on technological excellence often make the worst investment decisions.

To be successful, a high tech investor should have a broad understanding of how a company's products work and why those products are needed. It is also valuable to understand the company's place within its industry and to see the direction both the industry and the technology are taking. It can also be enormously profitable to know what role the company plays in determining the overall direction in which our society and economy are moving. The emphasis should be on conceptual understanding and a knowledge of a company's position in relation to technological change.

FOLLOWING UP ON THE INFORMATION IN THIS BOOK

It is very important for readers to know that the companies discussed in each section are companies to watch, not companies whose stocks you should rush out and purchase without further investigation. Many long shots are included, along with the industry leaders, if they have good potential and should be monitored for signs of progress. As is common with long shots, many will miss their mark.

Speculative companies and industry leaders alike should be thoroughly investigated before you make a stock purchase. I strongly suggest that you use the High Tech Directory, Part III, at the back of this book, and send for the annual reports of any companies under active consideration. In your request you might ask for reprints of articles that explain the company and its products. Most high tech firms are painfully aware that the majority of investors do not understand them or their business and are happy to send information wherever it will be put to good use.

Over the next several years, more people will make more money in technology investments than in any other field. Those who take the time to become knowledgeable about technology's many areas of growth will be miles ahead of those who don't and will reap the rewards of their foresight.

Part I

The Top Technology Investments

Section 1

A Revolution in Production Opens the Doors to Profit

CHAPTER 1

Industrial Automation: The Top-Growth Industry of Our Age

The Western world is again in transition. There can be little doubt that we are in the early stages of another industrial revolution whose effects on our society will be every bit as profound as were the effects of the first. Although the second industrial revolution is causing many disruptions, it is also offering the largest, most varied, and longest lasting investment opportunities of our time. Of course, as is true with most opportunities, the biggest profits will go to the best informed.

It is important to stress that the industrial automation movement is just beginning. Although some areas have received considerable publicity, only a very small percentage of any market for industrial automation products has yet been reached. The period of exploding growth is just now beginning and will continue through the turn of the century. I'm convinced that investors who take careful positions during the next year or so will very likely find themselves with the enormous profits that have accompanied every major development of modern times.

THE OLD AND THE NEW

The industrial automation movement goes back to the American Revolution, when Eli Whitney introduced the concept of standardized parts that could be produced in quantity. Twenty-five years later, Joseph Jacquard developed an automatic loom that was controlled by a set of punched paper cards. The invention of the steam engine by James Watt and the assembly line by Henry Ford have been followed by electronically controlled automatic machine tools. These developments have brought us a high degree of automation and in the process have made fortunes for millions of investors worldwide. But what we've seen so far pales in comparison to the technological advances and investment opportunities that are presently getting underway.

Although the factory automation movement has been on a plateau until the last few years, that plateau represents a high degree of sophistication. Anyone who has observed one of Detroit's automatic engine machines might wonder how it could possibly be improved. This enormous machine, which requires almost no assistance, takes in a large chunk of cast metal at one end and turns out a fully finished engine block at the other. Thousands of equally sophisticated machines exist throughout the world. But all such machines share one very significant limitation—they lack manufacturing flexibility. They are able to produce only one specific product and are cost effective only when producing large numbers of identical items.

However, our complex world is creating an overwhelming need for a wide variety of custom-designed products that must be produced quickly and efficiently in *small* numbers. Industries of every kind report that well over 50 percent of their orders are for less than 100 items. Those orders cannot be filled by the expensive, single-function, mass-production machines that are presently the mainstay of manufacturing industries. Limited quantities of customized items can be economically produced only by flexible systems that can turn them out at the touch of a button but with the efficiency of mass production. Fortunately for producers, consumers, and investors alike, advanced flexible manufacturing equipment and other essential systems have been developed that together can make up a fully automated factory.

THE ROLE OF PRODUCTIVITY

Increased productivity is clearly the moving force behind factory automation. Productivity plays a crucial role in determining how competitive and profitable an industry can be. The ability of a business to survive depends on the final cost of its product compared with that of the competition.

If you look at the costs that go into determining the final price of a product or service, you find there are really only five of them: the costs of money, raw materials, energy, capital goods, and human labor. Of these five basic costs, the first four are largely equal from industry to industry and between one country and another. Therefore, the only way a business or industry can increase its competitive edge is to decrease its human cost. That is, it must find ways to increase productivity. To this end, a factory must increase its use of high tech production equipment, for only such equipment can boost human effectiveness. The link between high technology, productivity, and profits is responsible for the incredible growth in the factory automation industry.

LOOKING AHEAD: INSIDE A FULLY AUTOMATED FACTORY

At the heart of the automated factory and each of its systems is a computer. Indeed, the design of the factory itself and the layout of its production machines is often the product of specialized computer systems that have been programmed to maximize productivity.

The fully automated factory consists of an automated warehouse that is the center of a computerized inventory control and material-handling system. Raw materials entering the factory are labeled, stored, distributed, and later shipped, all under the direction of a computer. All production is done by flexible manufacturing equipment serviced by robots that load and unload machines, then assemble the finished parts into the final product.

The company's products are not only manufactured automatically under the direction of a computer but are also designed with computer assistance. In addition, the computer translates the design into the detailed instructions needed by machine tools to actually manufacture each item.

Each system within a fully automated factory is under the direct supervision of a computer. The individual computers on the factory floor are connected to a master computer that oversees and coordinates the entire operation.

THE KEY TO SUCCESS

In the sections that follow, each part of the automated factory will be examined from the perspective of an investor in search of attractive opportunities. Each section is written with the conviction, proven many times over, that the key to making large profits in high tech areas is to have a basic understanding of the underlying technology, its contributions to productivity and products, and the direction in which each industry is moving.

Your competition in the investment arena is about to be outclassed.

CHAPTER 2

Booming Markets Develop for CAD/CAM and CAE

CAD/CAM means computer-aided design/computer-aided manufacturing. CAD/CAM has greatly increased the productivity of designers and engineers. Because the new technology has received so much publicity, many investors have mistakenly concluded that CAD/CAM has passed its period of greatest growth and is no longer attractive. In actuality, although growth may be cyclical the long term trend is upwards. Factors such as sharply lowered prices, increased capabilities, simplified commands, and the development of a whole new range of applications are expanding the market for such systems fivefold. However, many people simply haven't noticed.

The potential of CAD/CAM places it in the same category as robotics, automated material handling, automated manufacturing equipment, and almost every other system in the automated workplace. To best illustrate CAD and CAM, let us examine a specific example.

Fighter planes are an engineering horror of complexity. They contain miles of tubes, all of which are jammed into an incredibly small space and must conform perfectly to complex angles in order to fit properly. In the past, such tubes were designed and installed by a dozen people who needed six weeks to do the job and had an error rate of 100 tubes per aircraft. Now, a computer has been introduced to the project. The airplane's designs are stored in the computer's memory. The engineer merely selects the beginning and ending points for the needed tube and suggests a probable path. The computer actually designs the tube, checks it for errors, and then adds this new information to its memory.

The frosting on the cake is that, at the touch of a button, the computer can direct the tube's manufacture by translating the new design into a series of commands understood by an automatic bending machine. The result is a tremendous savings in time and greatly increased accuracy. It now takes only minutes to design and fabricate a new tube, with an error rate of only four per aircraft. The system vastly increases productivity and profits.

INSIDE CAD/CAM

The front end of CAD/CAM, computer-aided design, is a very advanced—almost intelligent—electronic drafting system. With CAD, a designer works on a special computer terminal called a graphic workstation, rather than on paper. Lines and angles are usually placed on the screen by selecting them from the computer's memory or by "drawing" on a special board with an inkless electronic "pen." The usual drafting tools are not needed because the computer will straighten lines, smooth curves, set sizes, and in every other way produce the finished drawing (however complicated) in moments. Because the computer can manage thousands of pieces of information with greater than human accuracy, the design will be error free and ready to use the first time.

The power of CAD to boost productivity is greatly increased when the system's computer is preloaded with specific information regarding the construction of a particular type of design. A program for designing factories, for example, will do large parts of the designing itself, such as working out beam placement, electric wire paths, or optimal machinery locations. Thousands of CAD programs exist for architecture, electronics, mechanical engineering, mapping, and many other applications not even dreamed of until recently.

Computer-aided manufacturing (CAM) picks up where CAD leaves off. It brings the power of the computer to the task of translating the design for a product into the product itself. For example, if an engineer designs a complicated part on a CAD system, he can then use a CAM program to determine the best way to make it. If the part is to be made with automatic manufacturing equipment, special CAM programs will establish the best machining procedures and even write the specific instructions for the computer-controlled production equipment.

Other CAM systems determine the best way to get the largest number of parts from the least amount of material. Programs have been developed that teach robots how to assemble products. Some programs even plan the most efficient order for manufacturing complex products and assist in directing the efficient flow of materials in a factory. CAM has not developed as quickly as CAD, but it is now beginning to catch up rapidly. The maturation of CAM will present many opportunities to knowledgeable investors.

There are three major groups of CAD/CAM systems. The first is a minimal-cost system that may sell for well under $25,000. Often these systems are dedicated to a single task and are ideal for users who work in one specialty such as semiconductor design or architecture. New systems for the IBM PC and AT are coming on the market almost weekly. The rapidly growing low-end market should be of prime interest to investors.

Next are systems that have more than one graphic workstation linked together in a cluster. Such systems begin at around $100,000 and are suited to the needs of medium-sized firms with complex design needs that require more power than the small systems offer. This market is also growing very rapidly.

The last group of CAD/CAM systems includes the largest and most expensive equipment. These systems allow many designers to work on different portions of one project by using separate workstations that are connected to the central computer, which updates and coordinates the whole design process. The high-end market is growing at a slower rate than those mentioned previously but it is still attractive.

THE GROWING ROLE FOR MECHANICAL CAE

Engineers need to do more than design products. They must also analyze the designs to make sure the products will perform as required. Many computer programs exist that can perform engineering analysis but because specifications must be entered one by one into the machine before calculations can be made, using them is painfully slow.

Not surprisingly, engineers have been badgering the CAD/CAM companies to add computer-aided engineering (CAE) to their systems so that once something has been designed on the computer, the analysis can be done directly. Engineers want the design to go smoothly from CAD to CAE and then on to CAM in a fully integrated system. CAD/CAM companies are moving quickly to satisfy the demand for CAE. Those that are first with top systems will have a distinct market advantage.

It is also becoming increasingly apparent that CAD/CAM companies have a market advantage if they buy their computers from established computer companies instead of designing and building their own. No CAD/CAM company can hope to supply computers that are as good as those that are available from IBM, DEC, and others who specialize in producing them. With the ready availability of inexpensive powerful computers, the CAD/CAM industry is rapidly becoming a software industry. The company with the best software is the company with the greatest potential. Be leery of firms that make all their own equipment.

A SHIFT TO COMPONENT SYSTEMS

More and more potential CAD/CAM customers already have computers. Such customers usually don't want to buy a complete CAD/CAM system whose built-in computer duplicates what they already own. They are more likely to want separate CAD/CAM components such as software and a graphic workstation that they can plug into their existing machine. Again we find that CAD/CAM companies with software developed to run on standard machines and a product line of individual components will have a competitive advantage.

The move toward component CAD/CAM systems is also being fueled by customers who have become more knowledgeable about computers. Such customers feel increasingly comfortable about designing their own custom CAD/CAM systems using components selected from several different sources. Unlike the automatic manufacturing equipment industry, which has few suppliers of component parts and almost no standards, the CAD/CAM industry permits considerable flexibility. Companies manufacturing graphic workstations, technical computers, and advanced software packages appear to be good investment opportunities because their long-term growth may exceed that of the turnkey CAD/CAM companies.

The rapid development of CAM and other factory automation equipment also encourages the development of component systems, especially those that conform to accepted standards. Factory automation is too complex a field for any one company to make all the necessary systems, from computers to robots. Because of this, a major effort is underway to build individual components that will work together. As an investor you should be alert to news about CAD/CAM firms that are stressing components that function with equipment from other parts of the factory automation industry. CAD/CAM companies that design their products to conform to emerging factory automation standards will prosper beyond expectations.

BUYOUTS ARE EXPECTED

As we've seen, CAD/CAM is an important part of the push to factory automation. Because of the size and complexity of the task, the movement favors the giant high tech companies that have the resources to tackle the job. Rather than lose time and market position reinventing needed technology, the big firms are buying out smaller high tech industry leaders. Any CAD/CAM company that is successful is a potential buyout candidate.

CAD/CAM is a foundation industry in the exploding factory automation movement. Knowledgeable individuals who understand the trends and forces occurring in the industry will be a giant step ahead of other investors when it is time to make profitable decisions.

Turnkey CAD/CAM Companies

Company	Symbol	Exchange
Auto-Trol Technology	ATTC	OTC

Comments: This company is a pioneer in CAD/CAM but fell well behind Computervision and Intergraph a few years ago. However, Auto-Trol shows signs of improved performance and may be a good turnaround prospect.

Company	Symbol	Exchange
Computervision	CVN	NYSE

Comments: Computervision has excellent products for almost every application. The company is now moving strongly into new areas, including semiconductor design. In April 1985, Computervision took a tumble in the stock market due to lowered earnings and unease among investors regarding the company's future. The company has probably lost market share to Intergraph but is strong enough to recover if properly managed. This company is a good turnaround candidate and should be followed.

Company	Symbol	Exchange
Intergraph	INGR	OTC

Comments: Intergraph is the world's top CAD/CAM company. The firm is in direct competition with Computervision and is gaining market share, probably at Computervision's expense. Intergraph has products for every part of the CAD/CAM market including inexpensive systems for the most rapidly growing applications. This company's long-range outlook appears to be excellent.

Component CAD/CAM Companies

Company	Symbol	Exchange
Adage	ADGE	OTC

Comments: Adage makes graphic workstations that enable a designer to interact with a host computer containing CAD/CAM software. High-speed fiber-optic links are now available for the most powerful applications. Adage is a leading maker of graphic workstations that are compatible with IBM mainframes.

Company	Symbol	Exchange
Apollo Computer	APCI	OTC

Comments: Apollo is the top maker of the supermicrocomputers that are widely used in CAD/CAM and CAE applications. This company is a top buyout or merger candidate. (See also the listing under computers, Section 4, Chapter 2.)

Company	Symbol	Exchange
Evans & Sutherland Computer	ESCC	OTC

Comments: Evans & Sutherland is one of the best computer graphics companies in the world. Although not a pureplay in CAD/CAM components, the company's workstations are widely used in high-end systems where top performance is needed. Other applications include simulation for flight training and computer graphic hardware for the motion picture industry. Good, long-range outlook.

Gerber Scientific	GRB	NYSE

Comments: Gerber's companies are involved in many levels of CAD/CAM. The firm makes full systems plus components, including large plotters for producing finished designs. Of special importance are the company's CAM cutters, which use plotter technology using knives, lasers, and other tools to create finished pieces on the factory floor.

Lexidata	LEXD	OTC

Comments: Lexidata competes directly with Adage in the market for high-performance graphic workstations used in CAD/CAM and other applications. The company's emphasis is on sales to original equipment manufacturers that incorporate Lexidata's workstations into their own systems. Definitely a company to watch.

MacNeal-Schwendler	MNS	AMEX

Comments: MacNeal-Schwendler is a leading producer of sophisticated software for computer-aided engineering. The company leases rather than sells its software, which ensures a continuing revenue stream. MacNeal-Schwendler is one of the very few companies with powerful mechanical CAE software.

PDA Engineering	PDAS	OTC

Comments: PDA also supplies software for designing and evaluating mechanical products. The software can display designs in three dimensions and in color, then generate a complete mathematical model that may be subjected to engineering tests. PDA went public in January 1985. Good growth since 1980.

CHAPTER 3

Robots Finally Come of Age, but Most Investors Aren't Noticing

Most of the machines you find in a factory perform tasks that humans are incapable of doing by hand, such as manufacturing parts with great precision. Since such machines do not replace humans, they do not need human capabilities. Robots, on the other hand, are designed to do what only humans can do. Many people failed to understand that distinction when the success of robots was first evaluated. It is very important for investors to realize that the success or failure of robots and the companies that make them depends on the extent to which the machines are able to function as substitutes for human workers.

THE PROMISE AND THE REALITY

Robotics is a classic example of a technology that was revealed long before it was able to fulfill its promise. Because the first robots were not able to replace humans, they failed miserably in most applications. However, they were glamorous and were the focus of an enormous amount of interest. Consequently, thousands of investors didn't notice the limitations of first-generation robots and suffered substantial losses.

As a result, discussing robotics is now taboo in many financial publications and robots are ignored by most investors. As one disgruntled trader mentioned to me recently, "I wouldn't notice a robot now if it walked down Wall Street playing 'The Star Spangled Banner' on a 12-string guitar."

However, robotics has matured over the past two years and robots are beginning to find places in modern factories. Many of the latest models are capable of doing unbelievable tasks and will be very profitable. However, due to the stigma surrounding robots, the new machines have not been widely noticed. Investors who understand this situation have an unusual opportunity to take careful positions in what will certainly become one of the all-time great, turnaround situations in technology.

The widespread use of robots is due to the industry's increasing success in solving the technical problems that have stymied it in the past. As an investor, it is very important that you understand the nature of these problems and their solutions, for two reasons. First, you will need to understand the background of robotics in order to avoid the investment mistakes of your predecessors and to see where the best investment opportunities are located. Second, as more of the technical problems are solved, your understanding of the significance of such developments will enable you to move quickly into the most profitable positions.

PROBLEMS, SOLUTIONS AND OPPORTUNITIES

Recently, important advances have been made in programming robots. Improvements in this critical area have had a great effect on sales. In the past, people on the factory floor who knew how to do the required work couldn't teach robots to do it because the machines understood only computer codes. As a result, programming and debugging needed to be done by engineers. The process was so slow that it was cost effective only for long-run operations such as automotive assembly lines. The biggest market for robots, medium-sized factories and short-run applications, couldn't be supplied.

Now, however, new-generation robots have programming devices that understand English, enabling them to be programmed by people without special training. Of even greater importance are robots that can be walked through a task by an operator, with all the movements recorded by the machine's memory for later duplication. When the latter technology is perfected, it will open many markets to robots and return attractive profits to those who produce them.

A major obstacle to placing robots in factories has been their inability to fit into existing operations. In the past, robots could only be used if an entire line was robotized. Because the cost of such large-scale systems was prohibitive, robot sales lagged.

Recently, however, the overall performance of robots has been improved. New, high-speed motors and computer controls are allowing the machines to fit into the normal flow of many factories. Now, one or two robots may be installed at an affordable cost, followed by additional units as finances permit. Performance advances are opening up markets rapidly, a process that will accelerate in the near future.

One way performance is being upgraded is by improving robot to robot communication. Humans do not work in isolation in a factory. They communicate with each other frequently so that the process of manufacturing or assembly may proceed smoothly. Likewise, a factory full of robots that are oblivious to each other is a factory that simply will not function properly. However, recent advances in high-speed data transfer technology have vastly improved communication.

Robots may be expected to more fully assume human roles in the factory as gripping devices (end effectors) continue to become more sophisticated. Although no such device can fully match the capabilities of the human hand, many advances have been made. Robots are now getting a sense of touch, mobility, and position awareness that was not available a year ago.

It is worth noting that in the area of artificial senses, U.S. companies are making enormous contributions and have a clear lead. You may have been surprised to see how many Japanese robot firms have formed partnerships with American companies to jointly produce and market complete systems. The Japanese, after all, are not known for sharing their business, especially in an industry in which the press would have you believe they hold a dominant position. In fact, these alliances have been hastily formed primarily to gain access to superior U.S. expertise in the essential area of robotic sensor technology.

ARTIFICIAL VISION, THE KEY ELEMENT

Significant advances in artificial vision have been made recently. These advances have enormous implications for the robotics industry. Before vision systems became available, robots didn't have the ability to adapt to the normal variations that occur in most industrial applications. If a part was upside down or wasn't in its correct position, the robot couldn't complete its task. Even the simplest variations often rendered robots totally useless.

Now, robots that can see and are affordable are coming on the market. They represent a totally new generation of machines that are no more like the first models than a new Porsche is like a Model T Ford. Thousands of industrial applications that were formerly closed to robots are opening up because of artificial vision. The two technologies, robotics and artificial vision, are a natural match and are beginning to generate significant profits. And it is occurring under the very noses of thousands of investors who gave up the robotics industry for lost.

The potential of artificial vision lies well beyond its robotic applications. Object recognition systems have great potential in such areas as parts inspection, inventory control, process control, machine tool monitoring, and security, to name only a few. Indeed, I predict that the infant artificial vision industry will soon develop into a separate area whose economic potential may rival that of the robotics industry. It is truly staggering to think of the possibilities that exist for machines that can see.

A major limiting factor in both robotics and artificial vision has recently been eliminated. A robot guided by an artificial vision system that requires several moments to analyze a picture before making a proper response would be totally useless in most applications. Fortunately, low-cost yet very powerful computers are now available that process data quickly enough to make these technologies useful in many new applications.

THE BENEFITS OF MERGING TECHNOLOGIES

You may have noticed that the revitalization of the robotics industry is the result of many different technologies coming together. If any critical part had been missing, the revitalization couldn't have occurred. Because the first generation of robots had only one of the necessary ingredients, sophisticated mechanical technology, it failed. Now that the necessary additional technologies have been developed, robots are finally becoming both useful and profitable.

IMMEDIATE APPLICATIONS

In addition to creating new applications, the increased capabilities of robots have led to their wider use in areas where they had already found a niche. Of particular importance is the movement into lower-cost markets whose total dollar value is huge.

For example, robots make good spray painters and welders, applications in which smooth, flowing motion is very useful. Painting and welding robots have been in automotive plants for several years. However, the new, easy-training systems are now making these robots cost effective in smaller industries as well. The new robots have proven extremely economical in dirty applications because companies employing robots for such tasks need not install the expensive ventilation systems and other safety devices that are required when humans do the work. In some installations, the net savings have been great enough to pay for the robots.

Some difficult welding applications formerly beyond the capabilities of robots have been opened up by artificial vision. For example, with even a low-cost sight system, arc-welding robots can now do an excellent job of following a continuous seam.

With their new capabilities, robots are becoming a very important part of flexible manufacturing systems that consist of a group of automatic machine tools working together to form a finished product. Robots load the tools, transfer parts from one machine to another, and sometimes assemble the finished products. Robots were formerly limited to very crude factory automation applications, but they are now found in semiconductor clean rooms making chips and in electronic factories assembling computers and other intricate products.

THE GROWTH WILL BE STAGGERING

In the United States today, there are something less than 20,000 robots installed, and that is a very liberal estimate. That number is less than the total number of workers employed in some individual factories. It is less than the population of hundreds of our junior colleges. The market is almost completely untapped.

Although the robotics industry got off to a disappointing start and has been plagued by many problems, the solutions to those problems are at hand and the industry presents extraordinary opportunities for profits. Even with their remaining limitations, robots are finding a growing number of applications in quite a number of industries. For the knowledgeable investor, the robotics industry holds great promise.

Robot and Artificial Vision Companies

Company	Symbol	Exchange
Apogee Robotics	APGE	OTC

Comments: This company is discussed in Chapter 5, the section on automated material handling, since that is the company's primary focus. Very promising.

Company	Symbol	Exchange
Automatix	AITX	OTC

Comments: This company is a leading artificial vision company for robotic and other applications. In the past, investors have been disappointed by stock price declines that followed unrealistic expectations about the speed the technology would develop, a familiar story with high tech firms. But the company is out of the primary development stage with products that work properly, and are selling. General Motors took a minority position in the company in 1984. Definitely a company to watch.

Company	Symbol	Exchange
Cincinnati Milacron	CMZ	NYSE

Comments: Cincinnati Milacron is not primarily a robot company, although it makes some of the best in the world. However, the company is a factory automation pureplay due to its heavy involvement with automated machine tools. See comments under machine tools, Chapter 4 of this section. Excellent company.

Company	Symbol	Exchange
International Robomation	ROBTC	OTC
Comments: This is a small robotics company with some capabilities in artificial vision. Garrett Corp. took a minority position in the firm in 1984. I think this firm is more speculative than Automatix or Robotic Vision Systems. However, the technology is too new to predict the eventual winners, so all participants should be reviewed prior to an investment.		
Object Recognition Sys.	ORSI	OTC
Comments: This company makes robotic grippers and robotic systems with vision components capable of selecting parts for assembly irrespective of their position. Good potential.		
Prab Robots	PRAB	OTC
Comments: Prab makes both material-handling and arc-welding robots, plus conveyor systems. Products are found in many industries, especially in heavy duty applications. If Prab adds sophisticated artificial vision systems it should increase its business significantly.		
Ransburg	RBG	ASE
Comments: Ransburg is heavily involved in worldwide sales of automated welding and spray painting systems, for which it is well respected. It is not a pureplay in robotics, but it will definitely participate in the factory automation movement.		
Robotic Vision Systems	ROBV	OTC
Comments: This is an artificial vision company that specializes in 3-D systems for guiding robots in many demanding applications. Like Automatix, with whom it competes, this firm has functioning products that sell. Potentially a very big winner.		

I've limited the preceding list to those companies that make robots primarily for their largest market, manufacturing applications. The list is also limited to companies for whom robotics makes a significant contribution to revenues.

CHAPTER 4

The Heart of the Factory: Automated Manufacturing Equipment

I've found from speaking to many professional investors and institutional fund managers that an unfortunate lack of understanding exists regarding automatic manufacturing equipment and its role in the factory automation movement. Unfortunately, much of the information available on this topic is of a general nature, and doesn't focus on the technology from an investor's perspective. Additionally, the role of automatic machine tools seems so obvious that it doesn't appear to warrant close inspection. Consequently, many people attempt to play this area by taking positions in a handful of old-line machine tool companies that don't have the proper technology to capture a significant share of the new markets. The result is almost always a loss of money and opportunities. Since there are enormous profits to be made in this field by informed investors, it certainly merits a thorough investigation.

AUTOMATIC ISN'T ENOUGH

All the various kinds of automatic manufacturing equipment share one characteristic: their movements are controlled by a set of premade instructions (programs). The earliest types date back as far as 1801. Their instructions were stored on rolls of punched paper such as that used by a player piano. Eventually, punched paper gave way to magnetic tape to hold codes for controlling the machine. Since these magnetic codes consist primarily of numbers, the machines that use them are said to be numerically controlled (NC). It is estimated that well over 100,000 NC machines are in daily operation across America.

Although numerically controlled machines have impressive capabilities, they also have several limitations that are not immediately apparent. First, most of them represent 20-year-old technology. Second, programming the machine is sufficiently difficult and expensive that they are usually cost effective only for large-run productions. Although programming NC machine tools has been simplified by using special computer programs offered by CAD/CAM companies, they are still not up to the needs of the automated factory. Finally, most numerically controlled machines are incapable of performing the wide variety of operations that are necessary for cost-effective manufacturing.

MICROPROCESSORS CHANGE EVERYTHING

Fortunately, numerically controlled machines have given way to a new generation of more sophisticated machines that are controlled directly by a computer. Computer numerically controlled machines (CNC) have many advantages over their predecessors. Programming CNC machines is simpler and faster than programming NC machines. Of equal importance, is the fact that a computer, with its built-in logic circuits and flexible response capabilities, can actually monitor work in progress and make appropriate adjustments as needed. Thus, dull tools, inadequate lubrication, and other problems are quickly spotted and corrected with CNC machines.

Perhaps the greatest benefit of all is that CNC manufacturing equipment can be linked together by a host computer into a coordinated direct numerical control (DNC) system. These systems offer the necessary ease and accuracy of programming and reprogramming that is essential for efficient, small-batch production—the major requirement of automated factories. In its most sophisticated form, a fully self-contained DNC installation is called a flexible manufacturing system (FMS) and is the most powerful factory automation product in existence.

If the master computer that controls and oversees the manufacturing equipment is connected to the company's CAD/CAM system that contains the design specifications for the products to be made, programming CNC and DNC machine tools can be all but automatic. Even in a less than optimal system where people still do the programming, that task is frequently possible using standard English. Incredibly, devices are now coming on the market that will allow automated machines to be programmed by voice.

MECHANICAL SUPERIORITY IS ALSO IMPORTANT

Significant advances in the technology of the machines themselves are of equal importance to the machine's computer links, a fact whose significance is all too frequently overlooked. Modern CNC machines are not simply old, single-function NC machines with a computer added. The best of the new machines are true machining centers where dozens of totally different operations may be performed.

A typical CNC machining center has a highly movable part-holding device surrounded by subunits that perform many different operations. The part-holding mechanism is almost robotic in its capabilities because it moves the part precisely from one machining subunit to another until all operations have been performed. When finished, a robot removes the finished product and replaces it with a new part to be machined.

Although a system of standard NC tools could be assembled to perform the same operations as one CNC machining center, in practice such a system wouldn't work as well and in the long run would not be cost effective. There is a market for less sophisticated setups, but the trend in the factory automation movement is clearly toward CNC and DNC machine tools, either working alone or, more commonly, in groups, to maximize productivity.

AVOIDING THE TRAPS

In summary, it is necessary not only that each automated machine be computer controlled, but also that the computers be compatible with each other and with a large host computer so that the machines may be linked together into a DNC or flexible manufacturing system. The machines themselves must be multipurpose machining centers and not simply upgraded and glorified single-function NC machines. It is important to keep the essential elements of a fully automated system in mind when making investment evaluations because the manufacturers of the more mundane NC machines are making a great public show about their new computer-controlled models. As we've seen, the computer is only the starting point.

Winning in the automated manufacturing equipment field presently requires that complete, ready-to-go (turnkey) systems must be supplied. Unlike the CAD/CAM industry, which has progressed toward component systems, the rest of the factory automation industry still lacks the standardization necessary for producing compatible components. In the past, customers have been disappointed by machines that couldn't work together. Now they want complete setups containing equipment that is fully compatible. The requirement for turnkey systems favors the largest suppliers, who have the resources to develop and integrate the various components. Beware of grandiose claims by small firms that they are successful suppliers of complete factory automation systems. Small firms simply can't do the job and aren't getting the orders.

SPACE-AGE MANUFACTURING

Many manufacturing operations that formerly used machine tools are increasingly employing new laser systems. Lasers can weld, cut, and polish with great precision and in many cases can do so for far less money than it would cost to use traditional machining methods. Laser beams are much more easily manipulated than heavy mechanical tools and, of course, they don't become dull and wear out with use. It is little wonder that in applications where laser machining is possible such systems are replacing the more expensive mechanical-machining centers.

Water jet cutting systems are also gaining market share in areas previously limited to more traditional and expensive manufacturing systems. Ultrahigh-pressure water jet cutting tools are frequently being used with fiberglass, corrugated box board, graphite, plastics, paper, and food products. In addition, new systems that use jets of water containing abrasives are being used to cut and polish many metals.

INVESTMENT TIMING

Sales of automated manufacturing equipment, like all capital expenditures, tend to be cyclical. However, there is considerable evidence that much of the new automated equipment will find markets even during business slowdowns because of its cost-cutting capabilities. Many industry leaders have confided privately that the lack of efficient production machinery was more devastating during the last recession than the direct effects of the recession itself. Too many firms went into the down cycle far too fat and inefficient. Those that survived learned their lesson and are not likely to let it happen again.

In many industries, automation is properly recognized as the key to survival. I expect recessionary fears will cause many companies to shift from equipment that increases production to equipment that simply cuts costs. Either way, the automated manufacturing industry wins.

Automated Manufacturing Companies

Company	Symbol	Exchange
Acme-Cleveland	AMT	NYSE

Comments: Acme-Cleveland is a leading machine tool manufacturer, but it has been slow to adopt computerized controls. The company is an excellent catch-up prospect and should be followed for progress. It wouldn't take much of a push for this firm to be propelled into the front lines of factory automation. I think the company will change, as the alternative is to drift into obscurity.

Company	Symbol	Exchange
AMCA International	AIL	TORONTO

Comments: AMCA purchased Giddings & Lewis, a leading maker of NC and CNC machining centers. AMCA is not a pureplay but its involvement is substantial.

Company	Symbol	Exchange
Boston Digital	BOST	OTC

Comments: Boston Digital makes ultrafine tolerance, CNC machining centers. The company is smaller than much of its competition but has good potential.

Company	Symbol	Exchange
Cincinnati Milacron	CMZ	NYSE

Comments: CMZ is the largest machine tool manufacturer in the United States. The company also is an important supplier of industrial robots and plastic-processing equipment. This is a top factory automation company capable of doing all but the biggest jobs. For the latter jobs, see Cross & Trecker, below.

Company	Symbol	Exchange
Coherent	COHR	OTC

Comments: Coherent makes several laser components and laser systems for industrial and other applications. Although the company is not a pureplay in factory automation it has a strong involvement and should benefit from the growth of the movement.

Company	Symbol	Exchange
Control Laser	CLSR	OTC

Comments: Control Laser produces several CNC machining and industrial marking systems that use lasers instead of cutting tools. The company is often able to do well during times when sales of more conventional machine tools are slow.

Company	Symbol	Exchange
Cross & Trecker	CTCO	OTC

Comments: This company is one of the few automated machine tool firms in the world that is capable of producing total, factorywide systems. Cross & Trecker is the company, largely overlooked in the United States, that the Japanese used as a model for factory automation. This company has yet to realize its potential but should do so soon. There is no one else in the United States capable of handling the big jobs which our basic heavy industries need to survive.

Company	Symbol	Exchange
Flow Systems	FLOW	OTC

Comments: Flow makes high-velocity water jet cutting systems and is considered the leader in its industry. The firm was formerly limited to applications involving nonmetallic materials, but it has recently developed a system that handles abrasive-laden fluids, opening up many new markets. Flow is a leader in an exciting technology with a bright future.

Company	Symbol	Exchange
G&B Automated Equipment	GB	TORONTO

Comments: This company is the world's leading producer of automated equipment for the bonded abrasives industry. G&B has considerable international exposure including sales to Soviet block countries (who pay their bills to ensure the continued flow of essential technology).

Company	Symbol	Exchange
Hurco Manufacturing	HURC	OTC

Comments: Hurco manufactures CNC metal-working equipment including gauging systems, vertical milling machines, and other automated devices. Although this small firm has experienced problems, it is worth investigating.

Company	Symbol	Exchange
JEC Lasers	JECL	OTC

Comments: JEC is basically in the same market as Control Laser. The company has several factory laser systems that have found good acceptance.

CHAPTER 5

Material-Handling Systems: The Tail that Wags the Dog

Automated material-handling and inventory control systems are often overlooked by investors in search of profits in the factory automation field. Yet such systems are an essential part of all modern factories and return good profits to their investors. Indeed, high tech material-handling companies did better than other factory system suppliers in the first phase of the most recent economic recovery. Although these systems may look like ugly ducklings, they are ugly ducklings that lay golden eggs. To my eyes, that makes them beautiful.

Automated material-handling systems are mandatory in modern manufacturing operations because an inefficient system simply can't keep up with high tech production equipment. Not only will poor material-handling equipment starve a flexible manufacturing system but it will also allow it to choke on its output. Because of the weak-link phenomenon, which allows an inefficient part of the productivity chain to limit the pace of the whole operation, it is necessary to have all the factory automation systems in balance.

In one well-known case where the various systems were not

balanced, $1 million worth of outdated production equipment was replaced by an $8 million fully computerized manufacturing cell. Everyone expected spectacular output, but the new equipment functioned no better than the previous production system. Of course the customer was less than overjoyed. However, once the material-handling system was upgraded, the factory functioned well beyond expectations and returned good profits to the company.

Speaking of profits, it is very important for investors to know that not only are automated material-handling systems and computer inventory controls a necessary part of the automated factory, but they are extremely cost effective. Factory managers and engineers in almost every industry report that a dollar spent on a high technology material-handling and inventory control system often pays a far greater return than a dollar spent on robotics or computer-automated machine tools. This undoubtedly accounts for the success of the material-handling industry.

UGLY DUCKLINGS AT WORK

As with every other system within an automated factory, each part of a material-handling system is linked to a network of computers. In the most advanced systems, the network includes a master computer that coordinates the individual units.

In order to keep the flow of materials and finished products moving properly, each computer in the network must have accurate information about everything happening within its jurisdiction. The system must know the location and the total count of every piece of material and every partially and fully completed product in the factory. This information may be fed into the computer in two ways. Workers can type it in using computer terminals scattered throughout the factory. In factories where there are few materials and parts and the manufacturing process is relatively slow, a manual entry system like this is adequate. The second method allows information to be entered into the computer automatically by remote sensors scattered throughout the workplace.

It has only been within the past four years that a simple and almost foolproof identification and tracking system has been available, but its presence in the workplace has revolutionized the materials-handling and inventory control industry. The system consists of bar codes and bar code readers whose growing importance to the field of factory automation, as well as to consumer retail businesses, has created a new growth industry that has considerable promise.

BAR CODES ON EVERYTHING

Bar codes are found on almost every product available today. The closely spaced stripes are arranged so that they will reflect a narrow beam of light and be converted into a string of coded pulses. This information is then fed into a computer that processes the data.

An industrial bar code system uses a precision laser scanning device capable of reading rapidly moving bar codes from a distance with amazing accuracy. In a recent test, bar codes were pasted on Frisbees that were then tossed back and forth through a new scanning system. The system recorded every one.

Bar codes are rapidly becoming an essential part of modern material-handling and inventory control systems. Raw materials are routinely bar coded upon entering a factory, then passed under a scanner so that a computer can add them to the total, direct their place of storage, and remember their location. In such a factory, a scanner is at every machining station, or every transfer point, so that the material is traced throughout the manufacturing process. It is also becoming standard procedure for all finished products to be bar coded and scanned so that the computer can arrange for warehousing, packaging, and shipping.

Bar code systems have an added benefit. They provide valuable feedback about the factory's performance that had been very difficult to obtain before their use. In many cases, information regarding performance is worth more than the direct increase in productivity that bar code systems bring to material handling and inventory control. Bottlenecks in the production process are quickly identified. Information about deficiencies in ordering materials, excess inventory, scrap rates, inefficient workers, poor shipping procedures, lost orders, and so on goes directly into the computer, thanks to bar codes.

Because bar code systems are such efficient information gatherers, a number of newly developed manufacturing computer systems utilize them in planning and controlling operations. Bar code technology is becoming so firmly established in the factory automation movement that it represents an excellent investment opportunity by itself.

Bar Code Companies

Company	Symbol	Exchange
Computer Identics	CIDN	OTC

Comments: Computer Identics manufactures scanners, decoders, and computerized processors for many applications.

Company	Symbol	Exchange
Intermec	INTR	OTC

Comments: Intermec is a leading supplier of scanners, scanning wands, bar code label printers, and related devices. The company is a pioneer in the industry and established several of the standard formats used today. A top company in the field.

Company	Symbol	Exchange
Scope	SCPE	OTC

Comments: Scope makes laser scanners for reading bar codes that are attached to fast-moving products. The company also makes material-handling systems for warehouse applications. This small company is rarely noticed but appears to have excellent potential.

Company	Symbol	Exchange
Symbol Technologies	SMBL	OTC

Comments: Symbol Tech makes a variety of bar code readers designed to connect with microcomputers. Because of the growth of microcomputers in small businesses, their use in low-cost bar code systems is becoming very popular. Symbol Tech should get increasing business from this application.

EFFICIENCY STARTS WITH INVENTORY CONTROL

Bar code technology, when used in conjunction with computer-controlled automatic storage and retrieval systems (AS/RS), can vastly increase efficiency in warehouses, which in turn can have a pronounced effect on the total productivity of a factory. A factory's warehouse is not just used to store finished products. It is also used to temporarily hold partially completed products and raw materials. Far from being nonessential, an efficient warehouse often sets the pace for the entire factory.

It is amazing the effect that even a minimal automation system can have on the operation of a warehouse. One of the simplest systems uses an inexpensive microcomputer tied to a network of lights located along the shelves. A worker can quickly locate items by entering their ID numbers into the microcomputer, which then activates the appropriate lights in front of the storage areas. Simple as they may be, systems of this type have turned warehouses that were horrors of confusion, requiring large staffs of experienced workers, into efficient operations requiring only a few unskilled employees.

The next level of sophistication involves linking a somewhat more powerful computer to an automated storage and retrieval system (AS/RS) that allows humans to be replaced altogether. Once the computer knows the proper location for all inventory, it is not especially difficult to send an automatic device to that spot. Many AS/RS systems employ automated carts with movable platforms that travel to the proper positions. Instead of turning on a light, the computer can then activate a mechanism that places the desired item on the cart or the shelf.

Other AS/RS systems resemble giant carousel trays such as those found on a slide projector. The storage device is rotated until the proper section is reached, at which time the desired item is dropped onto a conveyor system. For articles of relatively uniform size, very large AS/RS systems have been constructed that work like giant vending machines.

Whatever the form an AS/RS system may take, all have several things in common. They are run by computers that know what item is needed, how much there is of it, and its exact location. In addition, they are all highly efficient systems that require minimal human intervention. Lastly, they tend to pay for themselves very quickly.

FEEDING THE INDUSTRIAL ELEPHANT

Many low-tech factory distribution systems have been developed. An example is a towline system with carts that are pulled along using cables in the floor. Overhead monorail carriers are also very common. Such systems function well as links between manufacturing areas and automated warehouses, but only in factories that make large numbers of identical products such as automobiles.

But the factory of the future must have the capability of making small batches of different kinds of products just as efficiently as can presently be done with large numbers of identical items. For that reason, new distribution systems have been developed that are as flexible as the manufacturing and warehousing systems they serve. Automated guided-vehicle systems (AGVS) are gaining the most acceptance.

Automated guided vehicles are similar in some respects to towline systems, except that they run under their own power with the guidance of a computer. Automatic guided vehicles follow thin electrical wires or strips of fluorescent paint fixed to the factory floor. The vehicles come in all sizes and are capable of handling parts of every size and weight. If they are used in the presence of humans, they have safety systems that prevent accidental collisions.

When an automated guided vehicle reaches its destination, it may be loaded or unloaded by a robot or it may serve as the platform for the operation that is to be conducted at that point. The cart may leave the warehouse with an unfinished part that it then carries to a series of automated manufacturing machines. It may then deliver the finished product back to the warehouse. The automatic guided vehicle is doubly productive and cost efficient when it functions both as part of the distribution system and as part of the manufacturing system.

THE OUTLOOK IS GOOD

All the systems mentioned in this section are now in production and are selling well. Fortunately for investors, the vast majority of them are being produced by public companies. Since well over 95 percent of America's industries have yet to make the transition to badly needed automated systems, this industry is positioned for a long period of growth.

Automated Material-Handling Companies

Company	Symbol	Exchange
Apogee Robotics	APGE	OTC

Comments: Apogee makes automatic guided vehicle systems that move parts and materials in a variety of manufacturing industries. Of particular interest is the firm's increasing involvement in the electronics industry, which is likely to become a major buyer of automated equipment in the next several years.

Company	Symbol	Exchange
Engineered Systems & Dev.	ESD	AMEX

Comments: This company makes material-handling systems used in distribution centers and manufacturing applications. In addition, the company makes other factory automation systems including equipment for electronic factories. Definitely a company to watch.

Company	Symbol	Exchange
Harnischfeger	HPH	NYSE

Comments: HPH is not a pureplay in material-handling systems, but it is an important producer of automated warehouses and factory floor transport equipment. Because the company's other businesses, including heavy equipment and mining equipment, have not been strong in recent years, Harnischfeger's material-handling activity is having an unusually big impact on revenues.

Company	Symbol	Exchange
SI Handling Systems	SIHS	OTC

Comments: SI is a leading producer of several types of material-handling systems for manufacturers. In addition, the company makes automated order selection and inventory control systems for warehouses. SI has been in business for over 20 years and is well regarded in its industry. Well worth investigating.

Company	Symbol	Exchange
SPS Technologies	ST	NYSE

Comments: SPS created automated parts-handling and inventory control systems for its small parts customers. The business expanded quickly and the company now designs systems for many industries that must contend with huge numbers of small items. SPS has a growing list of electronic manufacturing customers.

Another public company that is a major factor in automated material handling is Eaton Corporation (ETN, NYSE). Unfortunately, Eaton is heavily diversified and, at the present time, can't be considered solely for its material-handling activities.

CHAPTER 6

Overlooked Profits in Process Control: The Other Half of Industrial Automation

It is all too common for investors looking for profits in the industrial automation movement to be distracted by all the publicity about robots, laser machining centers, and other spectacular factory automation equipment. As a result, many people fail to notice the more immediate and profitable investment opportunities in the process control industry, the other half of industrial automation.

Factory automation systems, such as those using automated machine tools, are designed for products that are manufactured one at a time. In contrast, industrial process controls regulate continuous or batch production as found in paper mills, sheet aluminum plants, or plastics factories. In these plants, sensors monitor production variables such as temperature, pressure, and thickness.

In most existing process plants, the information collected by sensors is displayed for human evaluation and response in what is known as open-loop systems. However it has recently become possible to plug sensors into reliable closed-loop systems in which computers, rather than people, are in control. Newly developed closed-loop systems are so cost efficient that switching over to them has progressed from being merely feasible to being mandatory for survival. Billions of dollars will be spent during this conversion period, with much of it flowing into the hands of knowledgeable investors.

DIFFERENT LEVELS OF CONTROL

There are three basic levels of process control. The first, simplest and still the most common, is the single loop, which consists of a sensor, a programmable controller, and a control instrument. With single-loop installations, one variable, such as the fluid height in a tank, is monitored and maintained. At this unsophisticated level of process control, closed-loop systems are more common than open-loop systems and frequently involve the use of inexpensive microprocessors.

Usually many single-loop systems are installed in a process facility. All of them report their status to a central location, where technicians watch the mass of gauges and try to keep the whole system functioning smoothly. Automatic coordination between loops is not yet widely used. Even in huge plants owned by companies whose names are familiar to everyone, the level of automatic control is low, and so is efficiency.

The second level of process control involves supervisory control systems that almost always contain a high-speed minicomputer. Supervisory control systems generally coordinate many single loops that together influence a major segment of the production process. For example, in a paper mill, a supervisory control system might oversee several loops that control the paper's thickness, its moisture content, its surface texture, and other physical qualities. The supervisory control system fine-tunes the process to maximize the efficiency of each segment of the production cycle.

Until rather recently, most supervisory control systems were designed by the firms that needed them. This was done dozens of times by rubber and steel companies. More recently, however, supervisory control installations have required the services of process control specialists who are as knowledgeable about computer systems as they are about sensors and instruments. As a result, most supervisory control systems are now designed by independent companies and installed as complete units.

At the top level of process control technology are millwide systems, still more a goal than a reality. At this maximum level of sophistication, a large computer oversees and coordinates many supervisory control subsystems or functions as a giant supervisory system itself. With millwide control, the entire process facility functions as a single entity. Such tasks as start-up and shutdown procedures, scheduling, and product alterations are greatly simplified by a millwide system—the Holy Grail of process control engineers.

Although many large process control installations have all their individual systems report to a central computer, in almost every case the computer's function is to inform engineers and management personnel about the status of the operation, rather than to actually control the plant. Such installations are frequently called millwide or plantwide management systems, which can be misleading. However, the value of having a single, shared database available to everyone in the plant is considerable, even if automatic control isn't part of the system.

INDUSTRY TRENDS AND INVESTMENT IMPLICATIONS

As is true in the factory automation industry, the process control field is clearly headed toward greater and greater integration of subsystems into true, millwide systems that function as single coordinated units. Although there are many small process control suppliers that make specialized products for specific applications, the greatest growth in this industry will almost certainly be among the market leaders that possess the resources to build large-scale systems.

Large companies have an additional advantage. The technology required to build specialized sensors and control instruments is now well understood. As a result, big process control companies are showing an increasing tendency to make such devices for themselves, rather than to obtain them from small suppliers.

Millwide system integration is as much a computer and data

network problem as it is a process control problem. As a result, future advances in process control will be marked by cooperative agreements with data processing and communication companies. News of such agreements should be of considerable value to knowledgeable investors who understand its significance.

Although it is obvious that process control firms with significant computer capabilities have a head start toward producing millwide systems, that lead is even greater than may be apparent. Linking individual supervisory systems into integrated units requires highly advanced software that takes years to develop and can be written only by people with considerable experience. Process control firms that have been slow to make the transition to computer technology will find themselves at a considerable disadvantage within a year or two. Purchasing powerful computers will not be enough to ensure development of powerful systems.

Because of the vast amount of data that must be collected and transmitted in large process control installations, high-capacity fiber-optic communication technology will shortly become mandatory. A single optical fiber stretched the length of a plant can carry more information reliably than a multiwire cable as big around as a man. Whichever process control company is the first to use this technology effectively will have a clear market advantage. None of the major process control firms have commercial fiber-optic systems ready to market, although some experimental ones have been installed. All the major firms are developing this capability as rapidly as possible. Investors should be alert for news about fiber-optic links in process control applications.

The process control industry presents very attractive long-term investments. It is an industry that is presently profiting from the first strong capital-spending cycle we've seen in many years. Because modern process control systems have the shortest payback period of any industrial automation investment, often under one year, they are being installed in record numbers by industries that know they must upgrade their facilities to survive. That represents a very large base of potential customers.

Process Control Companies

Company	Symbol	Exchange
AccuRay	ACRA	OTC

Comments: AccuRay is a top supplier of process control systems to the timber products, tobacco, plastics, and sheet metal industries. Paper production systems are its main focus. The company is very strong in Europe, as well as in the United States. AccuRay enjoys an excellent reputation for quality products, which undoubtedly accounts for the company's enviable growth record. The company is a leader in the supervisory control systems that are presently in great demand.

Company	Symbol	Exchange
Fischer & Porter	FP	ASE

Comments: F&P is a specialist in flow control systems. The company supplies process equipment to the petrochemical, food, metal production, and pharmaceutical industries. In addition, F&P's systems are used in water purification and sewage treatment plants.

Company	Symbol	Exchange
Foxboro	FOX	NYSE

Comments: Foxboro has a strong background in instrumentation and has become a leading process control supplier to a wide variety of industries. The company maintains over 175 offices in 80 countries. Foxboro's primary expertise is with loop control products, which it frequently ties together to form plantwide management systems.

Company	Symbol	Exchange
Honeywell	HON	NYSE

Comments: Honeywell supplies control systems to the process industry and to many other markets. The company's continued success in process control seems assured since Honeywell is also a major computer maker and can easily supply fully automated systems for the most demanding applications. Honeywell is in a very strong position to capture additional market share.

Company	Symbol	Exchange
Measurex	MX	NYSE

Comments: Measurex stands toe to toe with AccuRay in terms of the markets it serves and its technological orientation toward supervisory control installations. Both companies have their roots in sensor technology and were quick to realize that since they were the leaders in gathering data, they were in the best position to process it with computers. I am impressed with Measurex's Vision 2002 process control system, which has the capability of coordinating 12,000 loops into an integrated system that at least approaches the ideal of millwide control. Measurex appears to have a very bright future.

Section 2

Semiconductors: America's New Basic Industry

CHAPTER 7

The Dawn of the Age of Silicon

A few years ago, you could ask any observer of the American economy to identify our essential raw materials and basic industries and the person would mention iron ore, steel mills, rubber, and the like. However, as many workers and investors have painfully learned, our traditional industrial foundations have changed dramatically in recent years.

Today, silicon is perhaps America's most important raw material and semiconductor production is our basic industry. Indeed, semiconductors are rapidly becoming the brick and mortar of 20th-century society. Products that have a task to perform have been greatly improved by the sophisticated capabilities of semiconductors. Because the cost of silicon chips often adds little to a product's price, these chips are being adopted in unprecedented numbers.

Many people find it difficult and unsettling to be in the midst of a major technological transition period, but the rewards are great for those who understand and exploit the opportunities it presents. As we continue to move from an economy based on traditional heavy industries to one based on sophisticated technology, a simple change in focus is all that is needed to convert old losses into spectacular profits. The clear trend toward a semiconductor-based economy presents a multitude of opportunities for investors.

DIRECT AND INDIRECT INVESTMENTS

Taking direct positions in semiconductor companies can frequently be very profitable, although critics may point to the cyclical nature of the industry as a negative factor. Semiconductor suppliers will continue to be sensitive to economic trends, but they will probably be less so than in the past due to the growing importance of their products. In any event, the quite predictable performance of most semiconductor stocks may be used to good advantage during times of economic change.

Investors who are uncomfortable with cyclical events will also find profits in the semiconductor industry, but perhaps not with the better known firms. Newly developed custom chips produced by small specialized companies are in such great demand that sales are proving to be highly resistant to the normal swings of the business cycle. Orders for the new state-of-the-art chips will not only remain high, they should actually increase throughout the rest of this decade.

Very attractive investment opportunities also exist along the sidelines of the semiconductor industry with suppliers of production equipment, computer-based design tools, and automated testing devices. In fact, quite a number of sideline plays will almost certainly outperform most chip manufacturers in the next few years.

Overall, the semiconductor industry looks very attractive as a source of profitable long-term investments.

CHAPTER 8

A New Generation of Semiconductor Production Equipment Is Essential

The semiconductor industry has always been cyclical. During bad times orders fall off, only to pick up again as business improves. Chip makers expect the cycles and have always been able to plan accordingly.

However, the semiconductor industry was completely unprepared for the unprecedented demand stimulated by the recent economic recovery and the profusion of sophisticated electronic goods it generated. After experiencing slow to flat growth for three years, during which time capital spending for production equipment was all but nonexistent, the semiconductor industry was suddenly swamped with orders. So severe were the shortages of some chips that many products that relied on them couldn't be produced in anywhere near the quantities demanded by the marketplace.

Semiconductor manufacturers learned their lesson the hard way with unfilled orders and lost revenues. As a result, they started to invest heavily in badly needed production equipment. Although chip makers will continue to cut back on capital spending when business is slow, they are unlikely to severely curtail their present modernization programs.

NEW TECHNOLOGY ADDS TO THE DEMAND

A new generation of chips frequently referred to as very large scale integrated circuits (VLSIs) has contributed greatly to the need for increased capital spending for semiconductor production equipment. The new chips are as different from the old as an airplane is from a glider. They are too complex and too tightly packed to be manufactured using the production equipment that is presently the mainstay of the semiconductor industry. As a result, almost all chip makers require extensive retooling, which is extremely expensive and is expected to account for $3.5 billion of the $5.5 billion annual capital-spending tab anticipated for the semiconductor industry by 1987. Such tidy sums may be expected to warm the hearts and the pockets of knowledgeable investors who take the time to learn the basics of semiconductor production.

SEMICONDUCTOR MANUFACTURING

Unlike conductors such as metals, which allow electricity to pass through easily, and insulators such as glass, which block electricity, semiconductors permit limited amounts of electricity to pass through. More importantly, the electrical properties of semiconductors can be changed easily by adding small amounts of impurities. For example, a tiny silicon semiconductor with boron impurities on the inside and phosphorus impurities on the outside became the now famous electrical switch known as a transistor. We can take silicon and make a precision electronic device by carefully adding just the right amounts of specific impurities in just the right places.

The first transistors, which replaced vacuum tubes, were made one at a time from individual pieces of silicon, then connected together in circuits. Later, transistors were made several at a time on a single piece of silicon, then separated to be connected into circuits. Finally, transistors that were intended for a single electronic device were not only made together on one piece of silicon but they were also connected into integrated circuits (ICs) right on the silicon chip.

Small-scale integrated circuits soon led to large-scale integrated circuits (LSIs), and then to very large-scale integrated circuits (VLSIs). Of course, with each advance, the individual circuit elements became smaller, more tightly packed, and much more difficult to produce. Even the design for a very simple IC is so complicated it resembles a map of a large city.

In order to make the production of integrated circuits more economical, ICs are now produced many at a time on thin wafers of silicon that normally measure 5 inches across. Such wafers, which may well be the most perfect objects made by humans, are actually slices from cylinders of pure silicon crystal. Each wafer is divided into small squares called chips, each of which contains a separate IC. After fabrication, the individual chips are often tested while still part of the wafer. Those that work properly are separated, packaged, and usually retested.

INSIDE A SILICON WAFER

Integrated circuits contain thousands of elements arranged in many layers on a chip that is sometimes less than 1/4-inch square. The designs for ICs have become so complex that they must be created with the aid of computers.

Once a chip design has been completed, the computer transfers the plan for each layer into a pattern generator that uses a beam of light to draw the details onto special photographic film to form photomasks. These photomasks work just like projection slides. They are placed into a step-and-repeat camera that duplicates them many times, in rows and columns, onto larger photomasks, which are often the size of the wafers themselves. As you may have guessed, the large, wafer-sized photomasks are used to project the designs directly onto the wafers. The whole process is known as photolithography and is the foundation of semiconductor fabrication.

Transferring the designs for each chip onto the wafers is simple in concept, but the job requires equipment that is unbelievably precise. Bare silicon wafers are first treated with a material known as photoresist that is sensitive to light. A large photomask containing the design for one layer of the chip is then placed over the wafer. Ultraviolet light is focused through the photomask using a projection aligner, which transfers the image of the circuit designs onto the photoresist.

Wherever the light pattern projected through the photomask strikes the photoresist, the material hardens. As its name suggests, the hardened photoresist will not dissolve when washed with solvents and acids. However, the unexposed photoresist remains soft and is easily removed with chemicals, leaving the circuit pattern on the wafer.

The wafer is now etched by placing it in an acid bath, where the material under the hardened photoresist will be eroded away, leaving pits and/or channels in the exact pattern of the circuit design for that layer. In order to create the electrical properties that are needed for each layer, the wafers are exposed to selected impurities in a process known as "doping," then baked in a diffusion furnace to drive the impurities into the material.

The last step in this process is to cap each layer with an insulating material that is usually laid down by one of several thin-film deposition techniques. Again using photolithography and etching, "windows" are opened in the insulating cap where the circuits will connect with those on the next layer.

Each successive layer is formed using the methods just described until the entire structure is finished. The topmost layer is coated with a thin film of aluminum that will be the chip's electrical contact, after the unwanted areas are etched away.

NEW PROBLEMS AND SOLUTIONS

In World War II, intelligence agencies learned to transfer full pages of information via tiny photographs known as microdots, which were about the size of the period at the end of this sentence. The complexity and density of modern semiconductor circuits are hundreds of times finer than microdots. Features are so small that a single speck of dust could easily ruin a photomask or chip. As a result, the entire semiconductor production process is conducted under strict clean room conditions that make hospital environments look like desert sandstorms by comparison. Understandably, the market for clean room equipment is substantial. Since humans are the largest source of contamination in clean rooms, the market for automated wafer-handling equipment is also very large.

Circuit density is so great in the most advanced VLSIs that

light beams spread out too much to be used in making photomasks. Instead, electron beam lithography has been developed, which can make precise lines on the photomask less than a millionth of a meter wide. Electron beams may even be used to draw circuit patterns directly on the wafer, bypassing photomasks altogether. Although direct write lithography is presently quite slow, the technology is maturing rapidly and will be in full use within two years. Electron beam lithography has a bright future in VLSI production. In addition, X-ray lithography, which may achieve even finer resolution, should be ready for expanded use in a few years.

As serious photography enthusiasts know, the sharpness of a projected image falls off at the edges even when using a good lens. The same problem occurs when full-sized photomasks are projected onto wafers. Edge distortion makes such photomasks unsuitable for very finely detailed VLSI circuits. Instead, "wafer steppers" must individually project each design many times across the wafer in the same way step-and-repeat cameras create full-sized photomasks. Wafer steppers are mandatory for the complex VLSI semiconductors coming into common use.

VLSI technology has also made acid etching obsolete because this process erodes the tiny circuit lines and destroys sharp features. Dry, electrically charged gas from plasma etching or reactive-ion etching equipment is now used to produce more precise patterns in the wafer. Dry etching equipment has the added advantage of relieving semiconductor manufacturers of the environmental pollution problems resulting from the use of highly corrosive acids.

VLSI technology also requires specialized equipment for the doping process by which impurities are added to the etched wafers. The diffusion method doesn't offer enough control over how deep and how wide the impurities travel in the wafer, but with ion implantation, the impurities may be placed just where they are needed.

Finally, VLSI circuits need a wider variety of materials deposited as thin films than the old techniques can handle. Outdated methods have given way to "sputtering" systems that can place layers of complex compounds just a few molecules thick evenly over the surface of a wafer.

INVESTMENT OUTLOOK

I expect the semiconductor production equipment industry to grow a minimum of 25 percent a year overall through 1987, provided that reasonably favorable economic conditions prevail. Although the industry has appreciated substantially since the general market turnaround began in August 1982, I feel most investors have yet to grasp the full strength of the semiconductor industry's need for new production equipment. As a result, excellent opportunities for profits abound in this industry and will continue to become available to knowledgeable investors for several years. Because semiconductor companies and their suppliers are sensitive to business cycles, investors who are patient can expect to find outstanding values occurring periodically.

Semiconductor Production Equipment Companies

Company	Symbol	Exchange
Advanced Semiconductor	ASMIF	OTC

Comments: Advanced Semiconductor makes a broad range of wafer handling, processing, assembly, and packaging equipment, plus related materials. The company markets products worldwide and has been quite successful.

Company	Symbol	Exchange
Applied Materials	AMAT	OTC

Comments: Applied Materials is a leading supplier of thin-film and vapor deposition systems, which are used for the precise placement of substances on chips during manufacture. The company should continue to do well, both with its standard systems and with equipment for processing newly developed chips.

Company	Symbol	Exchange
Electro-Scientific	ESIO	OTC

Comments: This company is the leading supplier of laser systems for burning out bad circuit elements in chips. Once the defective elements are isolated, redundant elements can take over, resulting in a fully functioning product. Since lowering the scrap rate for chips boosts profits considerably, the company's products pay for themselves in a short time. Electro-Scientific has a growing market and should do very well as chips become more complex and low scrap rates more difficult to obtain. Definitely a company to watch.

Company	Symbol	Exchange
GCA	GCA	NYSE

Comments: GCA is a leading producer of wafer steppers and other wafer-processing systems. The company has a worldwide reputation for top products and service. In addition to semiconductor production equipment, GCA makes factory automation products, analytical instruments, and other devices. Excellent long-range outlook.

KLA Instruments	KLAC	OTC

Comments: KLA is a leader in automated optical inspection systems for semiconductor photomasks and other semiconductor applications. The company's products are particularly timely because structures in modern chips are becoming far too small for cost-effective inspection by humans using microscopes. Outlook appears very good.

Kulicke and Soffa	KLIC	OTC

Comments: K&S is a highly respected manufacturer of specialized devices that remove finished semiconductors from wafers and mount them in small packages that may be easily placed in electronic products. Not an overly crowded field, despite its importance.

LAM Research	LRCX	OTC

Comments: LAM is a small, young company that supplies the plasma-etching equipment that is used instead of acids to remove selected portions of semiconductors. Interestingly, while many U.S. semiconductor production equipment firms complain about Japanese competition, the Japanese have purchased up to 20 percent of Lam's output. Definitely a company to watch.

Materials Research	MTL	ASE

Comments: Materials Research competes with Applied Materials in the thin-film deposition segment of the industry. The company has been very successful.

Perkin-Elmer	PKN	NYSE

Comments: Perkin-Elmer is the world's largest supplier of semiconductor processing equipment. In addition, the company has the broadest line of equipment and is able to offer one-stop shopping to chip makers. More importantly, the company offers integrated products that work well together, which is an extremely powerful advantage in an industry where hodgepodge assembly lines are common. Of course, Perkin-Elmer is in the best position to carry integration to the next step, which is fully automated production facilities. Perkin-Elmer also makes analytical instruments and computers. Excellent long-range potential.

Company	Symbol	Exchange
Silicon Valley Group	SVGI	OTC

Comments: Silicon Valley makes specialized equipment for sawing, grinding, smoothing, and cleaning wafers, which are functions essential to successful chip production. In addition, the company makes equipment used in semiconductor photolithography. A top, front-end equipment company.

Company	Symbol	Exchange
Siltec	SLTC	OTC

Comments: Siltec is one of the world's few suppliers of silicon ingots and wafers, which are the raw materials of the semiconductor industry. In addition, the company makes equipment for handling, testing, and sorting wafers. The company should continue to do well during periods of increased semiconductor demand.

Company	Symbol	Exchange
Tylan	TYLN	OTC

Comments: Tylan has special expertise in the field of ultraprecision flow control. The company's products are used in diffusion furnaces and various types of thin-film deposition systems where the success or failure of the process depends on having total control over the amount and dispersal of the substances. The company has progressed from supplying components to making complete systems for semiconductor production.

Other public companies with significant involvements in this industry include: Plasma-Therm (PTIS, OTC, wafer etching), Machine Technology (MTEC, OTC, photolithography), Micro Mask (MCRO, OTC, photomasks), Nanometrics (NANO, OTC, wafer inspection), Optical Specialties (OSIX, OTC, wafer inspection), Sloan Technology (SLON, OTC, deposition equipment), and Veeco Instruments (VEE, NYSE, wafer etching and photoresist).

CHAPTER 9

Automating Semiconductor Design Breaks Production and Profit Barriers

Very large-scale integrated circuits are not only very difficult to produce, they are also very difficult to design. They have become so complex that creating the most advanced chips approaches the threshold of human ability.

One of the major obstacles to designing VLSIs is the difficulty of breaking down the task into small parts that can be given to a team of engineers to work on piece by piece. The problem is that even the smallest change in one part of the design affects every other part. As a result, VLSIs must be initially conceived in their entirety, although later in the design process they may be broken down into well-defined subsections.

Fortunately, a new set of computer and software tools has been developed over the past two years that is revolutionizing the process of engineering complex electronic circuits. Such tools automate the laborious and error-prone design of sophisticated semiconductors using techniques known collectively as computer-aided engineering, or CAE.

THE SECOND BIG GROWTH PHASE FOR CAE

The first generation of CAE tools for semiconductors was revolutionary and received much attention. However, the early devices were primitive compared to the powerful systems now becoming available. The CAE industry is starting to mature and reach its potential, but many investors haven't noticed.

Unlike the situation that occurred with lasers and robots, where people took positions too early and lost money, with CAE, most investors made initial profits and sold out well ahead of the large gains that lay ahead. Alas, not understanding the technologies resulted in lost opportunities in both situations.

The advantages that the latest CAE systems offer over conventional and first-generation CAE tools are so great that the industry that supplies them is presently growing 100 percent a year. Happily for investors, the top companies in the CAE industry are publicly held and are often undervalued when measured against their probable growth over the next several years. As is the case with companies that supply semiconductor production equipment, the semiconductor CAE industry represents an unusually attractive long-term investment.

COMPUTERS THAT CREATE COMPUTERS

Computer-aided engineering in its most basic form greatly simplifies and accelerates the design of semiconductors and other complex electronic devices. Instead of manually drawing symbols to represent the various electrical elements in a design—sometimes thousands of times, in a process that may take months—the engineer draws each element only once. When an element is needed again, the CAE system will call it up from its memory and place it correctly in the design.

But a CAE system isn't simply an electronic drafting tool. A CAE system actually assists in crating designs by monitoring each part of the structure for logical accuracy and violations of electrical rules. If a connection is missing or if an element can't function as specified, the CAE system will often detect the problem. After a design is completed, it may be tested in a process so thorough that in many cases, electrical prototypes need not be constructed.

The ultimate goal of the CAE industry is the development of a silicon compiler, a device that will accept a broad statement of desired functions and then automatically translate those functions into error-free production drawings. Although a full-range silicon compiler is still several years away, existing CAE systems represent an enormous step in the right direction.

RELIEVING A SHORTAGE OF ENGINEERS

Because CAE systems assist with much of the engineering that semiconductor designs require, the level of training needed to create excellent results with such systems is much less than is true with manual methods. Instead of needing the very special skills of one of the world's 2,000 fully qualified circuit engineers, sophisticated electronic devices may now be created by over 400,000 electronic engineers, a 200-fold increase.

The greatly expanded base of designers has an additional benefit to simply increasing the absolute number of semiconductors that can be created. CAE also permits companies to avoid the constraints of standard designs, allowing them to create and patent fully custom products with unique features that set them apart from their competition. Instead of becoming another "me too" company with a high likelihood of failure, electronic firms can diversify within a market and offer products that meet their customers' special needs.

CAE also helps companies deal with the crippling effects of accelerated technological change and rapid product obsolescence. Electronic products that formerly were marketable for four years or more may now only be useful for as little as two and a half years. The product development cycle must be shortened to match shorter product lives, all but mandating the use of CAE systems. This is especially true since manufacturing and testing, the only other parts of the production sequence, have already been automated. The last remaining bottleneck is the design phase and that can be broken only with CAE.

Prices for all electronic products are falling drastically, requiring manufacturers to reduce production costs wherever possible. Because CAE systems are even more cost-effective productivity tools than many types of automated manufacturing equipment, they are the first systems to be acquired.

Although CAE burst upon the scene so suddenly that many potential customers have yet to realize its full value, others are flocking to the systems. For both technical and economic reasons, CAE has become an essential part of the production process for advanced electronic products.

EMERGING TRENDS

The three major parts of the semiconductor production process are design, manufacturing, and testing. With CAE on the scene, the design portion of the sequence is becoming as productive as the other two steps. However there is one major additional opportunity for increased productivity and very handsome profits that is not apparent to people outside the semiconductor industry: the integration of design and testing.

It takes several months and a great deal of expensive talent to develop the software that is needed by the automatic test equipment (ATE) used to evaluate semiconductors. However, the software that is used in such instruments requires the same design information that is produced by the new CAE systems. As a result, CAE suppliers are finding that their products are even more valuable than they first realized. They are needed not only at the front end of chip production but at the final end as well.

The need for increased productivity in semiconductor production is so great that ATE companies absolutely must have access to the CAE systems that design the chips their products will test. The testing equipment companies can obtain the necessary technology in four ways. They can (1) enter the CAE market and attempt to beat the existing vendors at their own game; (2) form joint ventures or obtain licensing agreements with the existing CAE vendors; (3) buy out CAE companies; or (4) sit back and do nothing, in which case the CAE vendors will undoubtedly enter the testing equipment market themselves and give the ATE industry a run for its money. Not surprisingly, licensing agreements, joint ventures, and buyouts are becoming the norm, all of which can increase profits for CAE investors.

The growing CAE market is attracting competition from CAD/CAM companies, though the CAD/CAM firms will need to do a considerable amount of catching up. In any event, the market for CAE systems is expanding so rapidly that there should be increasing profits for the major vendors even though CAD/CAM firms are making an appearance.

As with the CAD/CAM industry, CAE products are becoming less expensive and more powerful, opening up huge new markets among small and medium-sized electronic firms. Some low-end systems are even based on IBM's AT personal computers and are beginning to sell extremely well. The development of the low-end markets is just getting underway and will bring millions in additional revenues to the CAE industry.

INVESTMENT STRATEGY

The CAE industry should remain attractive, due to rapid growth that will almost certainly continue for several years. However, a shakeout period common to all new, rapidly growing industries may be expected at some point in the future. As a result, I would expect greater than average price volatility to accompany the greater than average profit potential this industry offers. Investors should diversify within the industry and adjust their portfolios as the ultimate winners emerge.

Alternately, as we will see in the next section, investors may ignore CAE as a direct investment and instead take positions in custom chip firms that offer CAE primarily as a means to attract profitable semiconductor orders.

CAE Companies

Company	Symbol	Exchange
Daisy Systems	DAZY	OTC

Comments: Daisy is the oldest and most experienced of the publicly held CAE system companies. This company defined the CAE market and the direction for the entire industry. The workhorse of Daisy's line is a system that allows an engineer to design new systems, using any currently available chip technology, including gate arrays, standard cells, and full-custom circuits. Daisy makes its own hardware. Very good long-range potential.

Company	Symbol	Exchange
Mentor Graphics	MENT	OTC

Comments: Mentor bases its excellent CAE systems on computers made by Apollo. As a result, Mentor is able to commit its resources almost totally to CAE software development, which is proving to be a wise decision in this software-intensive industry. Because several of the company's officers once worked for Tektronix, a leading maker of test equipment, Mentor has been a leader in the integration of CAE and semiconductor testing. The company may find itself with two profit sources in a year or so as its involvement with test equipment begins to pay off.

Silvar-Lisco	SVRL	OTC

Comments: Silvar Lisco is unique among the CAE companies because it supplies only software. The company is in a good position to reach the growing market that already has CAD/CAM or CAE systems installed and simply needs more applications packages.

Valid Logic Systems	VLID	OTC

Comments: Since shipping its first system in May 1982, Valid has become one of the leading suppliers of CAE systems. The company's products utilize a software system with which an engineer can create full-custom VLSI circuits and semicustom logic circuits such as gate arrays or standard cells. To support its systems, Valid maintains an extensive computerized library of semiconductor designs and specifications from major chip manufacturers.

CHAPTER 10

Demand Explodes for Custom and Semicustom Chips

Closely related to the powerful CAE systems are customized semiconductors, which are often referred to as application specific integrated circuits or ASICs. Unlike standard off-the-shelf designs, customized chips are created for specific purposes. As a result, ASICs work fast and contain features unavailable anywhere else, which opens rich new markets to any capable electronic firm or chip producer with CAE design tools. Not surprisingly, custom chips have taken the electronic industry by storm and are becoming one of the fastest growing parts of the semiconductor industry.

The demand for ASICs is becoming so great that the firms that produce them have been much less affected by the cyclical business swings common to most of the semiconductor industry. During the slump in semiconductor orders that began in early 1985, several custom chip makers were bursting at the seams with orders while most producers of standard chips were shutting down production lines. At the present rate of growth, ASICs may account for 50 percent of semiconductor orders by 1990.

In order to make it possible for customers to more easily design ASICs, specialized chip makers offer their customers the use of their powerful CAE systems, which are optimized to produce customized chips. Quite a number of companies have established CAE design centers in various locations around the world to make it convenient for clients to create the products they need. In other cases, CAE capability may be made available by using direct communication links to the customer's computer. Of course, most makers of custom semiconductors will design chips for customers who are not comfortable doing the job themselves.

Pureplay CAE companies are also assisting custom chip makers by entering into mutually beneficial design-licensing agreements with them. In such agreements chip makers release their computerized libraries of standard cells and other designs to independent CAE firms who, in turn, make these libraries available to their customers. The CAE firms get a better product to offer at little or not cost, and, as you may expect, customers who make use of the licensed designs almost always have their chips produced by the firms that supplied them. Such ASIC business is especially profitable because the chip manufacturers did not have to make an expensive sales effort to get it.

PRODUCING ASIC CHIPS

The simplest way to obtain a semicustom chip is for the user to make alterations in standard semiconductors designed for a similar purpose. Certain "programmable logic devices" have their working elements equipped with built-in microscopic fuses. The devices are programmed to perform as desired by intentionally blowing selected fuses, leaving the remaining elements connected in the desired way. Programmable logic devices are the least complicated and least powerful semicustom chips. CAE systems are not necessary to design them, although special electronic tools are required.

The fastest growing of the more powerful semicustom devices are "gate arrays," which consist of standard masses of microscopic switches that are connected to perform as needed during the last part of the production process. Although they are not as efficient as full-custom semiconductors, they are easier and quicker to design. Because of their great popularity, design systems for gate arrays are made by all the top CAE firms, as well as by chip makers that specialize in them.

The first semicustom chip designed from the bottom up is composed of standard cells, which are small blocks of circuits designed to perform specific tasks. Large numbers of standard cells have been designed and are stored in the memory of a CAE system. Semiconductors are created by selecting and then combining standard cells that have the desired functions The leading CAE firms either have, or are about to introduce, standard-cell technology.

The most elegant and most difficult chip to create is a full-custom semiconductor. Full-custom ASICs are designed from individual circuit elements, not standard cells. These elements are combined with the aid of a CAE "cell compiler" to form efficient, compact structures. Cell compilers, more than any other CAE system, approach the industry goal of the fully developed silicon compiler mentioned earlier. As may be expected, cell compilers are more difficult to use than gate array and standard cell CAE systems, which has limited their acceptance. However, several of the leading CAE vendors are making progress with such systems and will be offering powerful products in the coming months.

INDUSTRY OUTLOOK

Custom and semicustom chips are produced primarily by a handful of companies that specialize in them. The more established semiconductor suppliers, whose mainstay has been standard designs produced in great quantities, have been slow to react to the demand for ASICs, although they are beginning to do so. However, the large semiconductor makers are not properly equipped to provide customer support and short production runs. At least for the next year or two the smaller, custom vendors should capture the lion's share of the business

Semiconductor distributors, on the other hand, have been very quick to react to customer demands for ASICs. Many of the top firms saw the potential of customized chips quite early but found themselves left out of the picture since customers needed to deal directly with the manufacturers to get their chips designed. In response, many distributors have been adding CAE design systems to their list of services and are now actively pursuing ASIC business. As may be expected, the custom chip makers have supported such moves wholeheartedly.

From every perspective, customized semiconductor manufacturers have an excellent chance to grow substantially. Not only do they represent a semiconductor investment with unusual potential, but they also provide a way to share in the growth of CAE systems.

Custom Semiconductor Companies

Company	Symbol	Exchange
LSI Logic	LLSI	OTC

Comments: LSI is a leading manufacturer of semicustom chips based on gate array technology. The company maintains full CAE services at several locations in order to simplify the customer's design process and shorten production cycles. LSI has been very successful, both in the United States and Europe.

Company	Symbol	Exchange
VLSI Technology	VLSI	OTC

Comments: VLSI Technology produces specialized CAE design tools for creating semicustom and full-custom chips. In the past, the company contracted with independent firms to manufacture chips designed on its systems. However, VLSI now has its own fabrication facility and offers design-to-product services. Because VLSI Tech's circuits are the most complex of the custom chips, the company has not enjoyed the explosive growth many investors once expected. However, the company's long-range potential appears very good.

Company	Symbol	Exchange
ZyMos	ZMOS	OTC

Comments: ZyMos manufactures semicustom chips based on standard-cell technology. As is true with other customer-designed chip makers, ZyMos offers CAE services. In the past, the company has not been as successful as some of its competition, but it should be reviewed prior to taking a position in this area. A rebound here could be very profitable.

CHAPTER 11

Automatic Test Equipment Completes the Picture

Once semiconductors have been designed and produced, they must be tested. For military and other critical applications each chip is tested individually, but more commonly a random sample from each batch is evaluated. In almost all cases, automatic test equipment (ATE) is used in an effort to increase speed and cut costs. Because ATE is very complex, there are only a few major suppliers, most of whom are public, which makes this area particularly attractive to investors.

TESTING IS ESSENTIAL

As I noted in the section on production equipment, semiconductor integrated circuits are made many at a time as individual chips on large wafers of silicon. Due to the delicate nature of the microscopic circuit elements, several chips on each wafer will usually be defective. For economic reasons as well as for purposes of quality control the defective chips must be identified so that after they are removed from the wafer, they won't be packaged. Chip evaluation is done with spiderlike probes connected to computerized test equipment and robotic wafer handlers.

All the functions of a complex chip can't possibly be tested. Instead, certain puzzles or conditions are imposed upon each chip, which reveal hidden defects. The process is analogous to the standard technique for finding a flaw in a crystal goblet. Instead of examining each goblet inch by inch in a process that could take several minutes, goblets are simply tapped to create a sound. Goblets with tiny cracks or bubbles make a dull sound, whereas perfect goblets ring like bells. Semiconductor testing depends on equally clever procedures that are programmed into the software used by the equipment. As is true of bad goblets, defective chips usually can't be repaired and are discarded.

Chips that pass initial tests done on the wafer are separated and mounted in plastic packages suitable for placement on circuit boards. In most cases the packaged chips are again tested before being shipped to the customer. Frequently, customers will test the semiconductors one final time before mounting them. Fully assembled circuit boards are themselves tested using products designed by another segment of the ATE industry.

Testing must be extremely thorough since a modern electronic device sometimes uses many thousands of semiconductors. If even a small percentage of defective chips were permitted, almost no finished product as complex as a computer would work. Individual ATE systems for semiconductors are priced from approximately $100,000 to over $1.5 million.

DETECTING SPECIAL CHIPS

In addition to providing quality control capabilities, ATE products are often employed to locate those chips that function much better than normal and those that function below standard. Top-performing semiconductors command a premium price for military, medical, and certain computer applications, and are well worth selecting for those special markets. Since substandard, but not defective, chips may frequently be used in a less demanding product than they were designed for, they may also have value. In fact, substandard but otherwise very powerful chips that were intended for computers and other sophisticated devices often end up in electronic toys. Readers may recall the numerous news photos seen every

Christmas of smiling Russian diplomats putting bags and bags of talking dolls and plastic robots on Aeroflot flights to Moscow, "for the kids at home." You may be certain that few, if any, of those toys end up in the hands of Soviet children.

INDUSTRY TRENDS AND INVESTMENT IMPLICATIONS

As semiconductors become more complex, the need for testing will increase, further improving the fortunes of the relative handful of ATE suppliers. In addition, new machines will be much more sophisticated and expensive than those they replace.

The growth of semicustom and full-custom chips will also place upward pressures on the ATE industry. If, as is expected, ASICs make up 50 percent of the semiconductor industry's output by 1990, approximately 30 percent of the existing base of ATE will also need to be replaced. That percentage may be low, as it is calculated without adding the expected growth in the overall number of chips produced.

The growth of military markets for semiconductors also favors the ATE industry, since all Department of Defense contracts provide for strict testing. Due to well-publicized problems with insufficiently tested chips from major companies, military testing requirements have been stiffened and are being more rigorously enforced. ATE companies are crying crocodile tears for their unfortunate customers because the new rules are boosting sales.

ATE makers will continue to seek links to the developing CAE industry. At this time, it is difficult to predict which additional companies will form alliances, although it is safe to assume that most will do so. However, since most ATE firms are much larger than their CAE counterparts, it is easy to see which is most likely to be Jonah and which the whale. If you are interested in taking a position in a buyout candidate, stick with a CAE firm.

The semiconductor ATE market is primarily shared by Fairchild (a division of Schlumberger) and the companies listed in the accompanying table. Although the top Japanese firm in the semiconductor ATE industry aggressively sells systems in the United States, their shares are not traded on U.S. exchanges.

However, they may be obtained by U.S. brokers for individuals or institutions who intend to hold them for long-term appreciation, thereby justifying the somewhat greater costs and delays involved.

Semiconductor ATE Companies

Company	Symbol	Exchange
LTX	LTXX	OTC

Comments: LTX specializes in producing test equipment for linear and linear/digital semiconductors, the types of chips often encountered in telecommunications, automotive controls, and consumer electronics. The development of very large-scale integrated circuits (VLSIs) for the latter markets will add to the demand for new testing equipment from LTX. The company competes with Teradyne but has a narrower focus. LTX should continue to profit with the growth of the semiconductor industry.

Company	Symbol	Exchange
Reliability	REAL	OTC

Comments: Reliability makes specialized "burn-in" equipment, which subjects semiconductors to the conditions they encounter in actual service. Usually, defective chips will fail in the first few hours of a burn-in test. As a result, chips designed for military and other critical uses almost always go through such a test before being certified. As semiconductors become more complex and we become more dependent on them, revenues at Reliability, Inc. should continue to grow.

Company	Symbol	Exchange
Takeda Riken	—	TOKYO

Comments: Takeda is almost unknown to U.S. investors, but it is familiar to semiconductor manufacturers both here and around the world. The company is the largest supplier of LSI and VLSI test systems in Japan and may well take the number two spot in the world. Takeda's systems are fully integrated and communicate by means of high-speed networks. The company is a leading ATE supplier with a promising future. Shares (via ADRs) are available in the United States through Merrill Lynch, Nomura, and other leading firms.

Company	Symbol	Exchange
Teradyne	TER	NYSE

Comments: Teradyne is a top supplier of ATE equipment for semiconductors and other devices. The company invested heavily in R&D over the past few years, which is paying off with many advanced devices for testing the new VLSI chips. One new product can test several chips at once and is breaking bottlenecks in testing lines for an increasing number of manufacturers. The company has a much broader market than LTX. The long-term outlook appears excellent for Teradyne.

Section 3

Communication: Still the Key to Progress

CHAPTER 12

From Tomtoms to Telecom

Virtually every advance in the human condition depends on effective communication between people. While individuals in every field are often able to make important breakthroughs acting alone, the discoveries they make cannot be put to effective use without the involvement of many others.

Communication is actually more important than may first appear. Not only does transmitting information permit the exchange of ideas, but it also makes it possible for many people to organize themselves to attain almost unbelievable goals. Andrew Carnegie summed it up many years ago when he coined the term *mastermind principal* to describe the phenomenon of creating a superbeing out of the well coordinated efforts of several talented individuals.

When we look at the pivotal role that communication plays in our lives, we can see that it is no accident that the progress of the human race closely parallels progress in communications. Neither is it any wonder that of all our technologies, communications receives our greatest emphasis. To understand our dependence on communication is to understand the great potential the field holds for knowledgeable investors.

COMMUNICATION GOES DIGITAL

We have been working on communications between humans for several years. We've managed to make good progress with technologies that transmit written words and voices. However, over the past 20 years or so we have created a major problem for ourselves by becoming highly dependent on electronic devices to help us think and do much of our physical work.

Unfortunately, our new electronic assistants use digital signals instead of voices, so they can't communicate efficiently using the bulk of our existing telecommunication technology. As a result, new digital communication equipment is coming into widespread use throughout the modern world. The adoption of the new devices is occurring at breakneck speed and will ultimately cost billions of dollars, much of which will find its way into the pockets of investors who take the trouble to see the trends and plan accordingly.

THREE AREAS EXPAND RAPIDLY

Within an organization, digital devices are being linked with local area networks (LANs) nearly as fast as the new links can be produced. In offices and on the factory floor, electronic equipment of every type is being transformed from independent devices to coordinated systems. LANs are rapidly becoming one of the major growth technologies of our age.

In situations where both people and digital devices must communicate using the same system, modern private branch exchanges (PBXs) are finding ready acceptance. The new PBX systems are not only effective but also very cost efficient, which is causing sales to accelerate.

Across long distances, several types of digital communication equipment are required ranging from small coding devices for computers to satellites. Such equipment is needed in unprecedented quantities as business communication needs continue to expand.

Over the short term, no investment, including an investment in communications, can be counted on to perform as desired. But over the longer term it is difficult to imagine a class of investments with a more likely chance to appreciate substantially. Digital communication is becoming especially critical to human progress and has a very bright future.

CHAPTER 13

Office Telephones Take on New Roles and Mass Installations Begin

Everyone is familiar with the typical office telephone system which is, in many respects, a miniature version of the national system. All calls go through one central telephone or small switchboard, which routes them automatically.

Several years ago, top-of-the-line office telephones, known as private branch exchanges or PBXs, began to acquire expanded capabilities such as internal call forwarding, electronic message taking, and other amenities. That trend toward greater power is now accelerating and is reaching the point where office telephones are rapidly becoming the central nervous systems for businesses. Because modern office telephones are easy to use and are very cost effective, they are being adopted in numbers reminiscent of the early years of the computer revolution and will put millions of dollars in investors' pockets.

MICROCHIPS EXPAND CAPABILITIES

The once humble PBX has recently acquired computer capabilities and is rapidly becoming the most essential part of the modern, automated office. In their present form, PBXs serve as central connecting and coordinating systems for all the individual parts of the electronic office, including personal computers, telephones, facsimile machines, word processors, and even slow-scan TV. PBXs with "stored program control" automatically route information from each source to its correct destination, whether that is across the office or across the world. Of course, in addition to its computerized heart, the key to the PBX's success is its ability to handle both voice and data communications with equal efficiency.

COMBINING VOICE AND DATA

The human voice makes very inefficient use of communication systems. Our speech, with all its melodious tones, is transmitted over telephone lines as a continuous wave, called an analog signal, most of which is a waste of system capacity. Compared to data transmission, speech conveys very little information for the amount of time it requires.

On the other hand, computers, word processors, and other office automation machines containing microprocessors communicate by means of digital signals that consist of simple impulses used to represent data. Since digital signals are much more compact than analog signals, they move a great deal more information in a given amount of time than is possible by voice. For example, this book, which would take hours to read over the phone, could be transmitted electronically in just a few minutes by most digital communication systems.

Although the most efficient and economical way to transmit information is to use digital signals rather than voice most of the nation's telephone systems were set up to transmit only the narrow range of frequencies found in analog signals. The telephone system lacks sufficient space for broadband digital transmissions.

It is possible to send digital information over standard telephone lines but in order to do so, the digital signals must be given a "voice" by converting them to tones using a device known as a modem. After the sounds representing data reach their destination, they are converted back into digital signals by another modem. The process is far slower than if digital signals were sent over special digital lines called T-carriers. However, until more digital lines are installed by AT&T and other long-line services, most PBX systems will still contain modems for outside computer to computer communications.

Within the office, however, digital information from computers and other equipment goes back and forth unchanged because modern PBXs are themselves digital systems. The new digital PBXs, which are the heart and soul of the office telephone revolution, transmit voices by first converting them into efficient digital signals using a codec. Many telephone conversations can then be combined using a multiplexer and carried simultaneously by the system. In addition, with a purely digital system, one set of wires can be used for all types of communications.

The efficiency of a modern PBX system extends beyond the office. When an outside communication is needed, many PBXs quickly select the best and least expensive line based on the type of communication to be sent and the rates charged by the various long-line companies for that particular time of day. If a T-carrier is available, outgoing digital signals may be combined so that they will take a minimum amount of time for transmission. If not, costs can still be kept under control by using high-speed modems to transmit information.

THE EXPANDING PBX MARKET

President Carter's decision in 1978 to end the American Telephone & Telegraph Company's (AT&T) monopoly on telephone equipment opened an enormous market for PBXs by making it possible for independent suppliers to sell systems to businesses. Of equal importance is the more recent divestiture of AT&T's local telephone companies, which are now free to purchase their PBX equipment from anyone they choose. The seven new Regional Bell Operating Companies (RBOCs) represent a major new market for PBX equipment, with sales expected to quadruple by 1987. AT&T, of course, will compete with other companies for PBX sales to both the RBOCs and businesses. However, much of AT&T's equipment lacks the sophistication and the cost advantages of systems available from independent sources and is not likely to capture most of the market.

The forerunner to today's PBX is AT&T's Centrex System, which is in use by thousands of businesses and is now considered out-of-date. Although the RBOCs are attempting to keep Centrex alive, it is almost certain that at least half of these systems will be replaced in the next few years.

The market for PBX systems is also expanding in response to the growing amount of office automation equipment that is becoming available. Businesses that would ordinarily find it possible to get along with more conventional electronic telephone switchboards are now finding that they need a more powerful digital PBX to pull all their office systems together to maximize efficiency.

Companies with growing digital communication needs have the option of installing Local Area Networks (LANs) instead of PBXs for computer to computer communications. For many applications, however, a digital PBX is just as effective and provides top-notch telephone service as well. Although LANs are needed for high-volume digital transmissions and, as we will see in the next chapter, are of growing importance, modern PBXs are adequate for many office applications. Businesses that do require LANs will soon have several new PBX systems to choose from that have LAN connections.

The increasing number of digital T-carriers being installed by long-line suppliers is increasing the utility of PBX systems. Many businesses with old analog-type PBX equipment will upgrade to digital systems as the new lines become available.

The growing number of alliances between PBX and computer companies is having a major influence on sales. Businesses that were unsure what role a PBX system would have in the integrated office now know such systems are an integral part of the total picture. IBM's takeover of ROLM Corporation was a validation of the entire PBX industry and dissolved an important barrier to acceptance. In addition, these alliances permit customers to obtain completely integrated turnkey office automation systems supplied from a single source.

The emergence of alternative long-distance carriers that offer rates well below those of AT&T is also fueling demand for PBX equipment. The previously mentioned capability of many digital PBX systems to select the most cost-effective carrier makes it possible to calculate a system's payback period quite accurately. In many cases, payback occurs within two to three years, which greatly increases sales.

An embryonic but potentially large market for PBX equipment is in new commercial buildings. Multi-tenant systems place the power and convenience of full-function PBX equipment at the disposal of businesses that would ordinarily be too small to justify its purchase. By including such systems in new buildings, real estate developers are finding that they can significantly increase the appeal of their holdings.

Modern digital PBXs are multipurpose communication devices that increase efficiency and often pay for themselves very quickly. It is little wonder that they are becoming one of the fastest growing segments of the telecommunications industry.

COMPUTERPHONES MAY FOLLOW

Hybrid PBX and computer systems are coming to market this year, which may provide unexpected revenues to most PBX makers and a few computer firms who are their main suppliers. The new devices, known as computerphones, take the digital capabilities of a PBX one step further by substituting a full microcomputer for the microprocessor. Of course, capabilities are enhanced considerably.

After getting off to a slow start, computerphones are beginning to make enough headway that the larger suppliers are planning to push them aggressively in the coming months. If they are successful, the telephone companies have expressed an interest in supplying them. Should that occur, computerphone sales could reach $200 million by late 1986 and perhaps as much as $1 billion by 1990.

I urge you to monitor the development of computerphone sales for signs of significant progress. Computerphones have the potential of having a big impact on the fortunes of several PBX and computer firms.

PBX Companies

Company	Symbol	Exchange
Compaq Computer	CMPQ	OTC

Comments: Compaq is the leading supplier of IBM-compatible microcomputers. In late March 1985, the company announced a new line of computerphones, which it refers to as Telecomputers. By the time this book is published the new machines should have been out long enough for investors to make a reasonable judgment about their chances for success. Of course, Compaq is no pureplay, but if the Telecomputers sell well it should have a big effect on the company's revenues.

Company	Symbol	Exchange
InteCom	INCM	OTC

Comments: InteCom is probably the top supplier of PBXs for large businesses. Many installations include 2,000 phones and up. InteCom's customer list reads like "Who's Who in American Industry." This is due in part to the versatility of the company's systems which can be programmed and reprogrammed by customers to fit their changing needs. The company is also a leader in multi-tenant PBX systems. Outlook is good.

Mitel	MLT	NYSE

Comments: Mitel lost out to ROLM in its effort to supply IBM with PBX systems. However, the company is still an important supplier in some markets and has the potential to become stronger in the future. Don't make a PBX investment without checking Mitel's current status.

NEC	NIPNY	NYSE

Comments: NEC is not a pureplay in PBX systems but is fast becoming a very important supplier and will undoubtedly give North American firms tough competition. The company is a leading maker of computer and telecommunication products and is therefore well equipped to supply computerphones if the demand for them increases.

Northern Telecom	NT	NYSE

Comments: NT is the acknowledged leader in digital PBX technology and is also well positioned to benefit from an increase in demand for computerphones. From its inception, the company's products were designed to be expandable and in many cases upgradable. As a result, the company will undoubtedly generate substantial revenues from its huge base of existing customers, in addition to supplying new business.

CHAPTER 14

Local Area Networks Fulfill the Promise of Computers

All of us have seen many glowing reports about the advantages of the computer revolution, with computers shown working together harmoniously in almost every industry to "free humankind from dull or dangerous" work. Unfortunately, the reports describe the potential of the computer revolution, not its present condition.

The truth is that a majority of businesses and industries have found that the new electronic technologies do not fully live up to their expectations. Many of the advantages of modern office and factory automation devices don't occur unless the individual elements are joined together into coordinated systems. Unfortunately, the means to do so has been elusive. Until recently, the necessary connections, known as local area networks (LANs), were woefully inadequate and contributed very little toward alleviating the pent-up demand for integrated computer systems.

In the past few months, however, extremely sophisticated and yet affordable LANs have been developed that begin to fulfill the promise of the computer revolution. The new networks are finding ready markets among computer users in every field who have been awaiting their arrival for years. The widespread adoption of LANs is just now getting underway and will certainly be the next big movement in electronic automation.

The potential size of the growing LAN market is staggering. Dataquest recently reported that LAN sales will be up 56 percent this year, to $224 million. The company expects the market to reach $1 billion by 1988.

LANs COMPLETE THE AUTOMATION PICTURE

One of the biggest advantages offered by a group of computers that have been joined together with a LAN is that everyone in an office, or every computer-controlled machine in a factory, can share the same information, significantly increasing efficiency and productivity. Changes made in one part of the system are immediately available to every other part of the system. No one lacks up-to-the-minute data needed for all parts of the job. Users report that LANs greatly reduce errors and the amount of time lost due to poor communication between team members, whether human or electronic.

Of equal importance to the contributions of LANs in existing applications is their ability to make possible new projects that were formerly beyond reach. The design and manufacture of complex semiconductors, for example, requires the use of several types of computer-based devices that must frequently exchange massive amounts of information at high speeds. It is no coincidence that advances in semiconductor design and many other technologies follow closely behind advances in LANs.

In many situations, LANs pay for themselves very quickly by greatly reducing the need to duplicate expensive peripheral equipment. Laser printers, plotters, data storage units, and other devices may be shared by many users if they are part of a local area network.

LANs with the most appeal are designed to link equipment from many different suppliers. By choosing general purpose LANs from independent companies rather than brand-specific LANs from computer firms, customers can mix and match equipment from various sources to suit their particular needs. Customers particularly like not being tied to any single supplier for necessary technology because no one company makes everything.

LANs ARE NO LONGER A LUXURY

The need to install effective LANs is rapidly reaching critical proportions, due to the proliferation of high tech office and factory equipment that has occurred in recent years. Traditional mainframe-to-terminal links weren't designed for the many types of graphic workstations, word processors, CAD/CAM equipment, satellite transmission systems, and other sophisticated devices that are now in common use.

Due to a lack of effective device-to-device communications, many newly developed and badly needed products are limited to only a few applications. We've reached the point in many areas where the major obstacle to progress isn't the lack of more powerful machines but the lack of an effective means to link existing equipment into coordinated systems. Newly developed LANs have become essential.

The widespread adoption of personal computers (PCs) is also contributing strongly to the need for LANs. Almost overnight, PCs replaced terminals as the primary workstations in most offices. It is vitally important for the efficient operation of thousands of companies that PCs be brought into the mainstream of office automation systems. Fortunately, several new LANs give PCs the necessary connections to other PCs, large computers, high-speed printers, and other types of office equipment.

WHY IT TOOK SO LONG

It is not at all difficult to transfer data from one piece of electronic equipment to another. All that is needed in the simplest case is a set of wires plus a network controller board in each machine to serve as a liaison between the network and the equipment. The Tough Part—and that is Tough with a capital "T"—is making the data understandable when it arrives at its destination.

The situation is analogous to the telephone network. With international direct dialing, it is easy to call someone in the Far East or Europe, but getting your message understood is another matter. With computer-based equipment, the problem is compounded because different devices use totally different instruction codes (software), in addition to different languages. Look how long it took to get integrated software for microcomputers that makes it possible to share data between different parts of the same program within the same machine. Add totally different machines to the picture and the scope of the problem confronting network engineers becomes more apparent.

AN ELECTRONIC PARTY LINE

Moving data within a network and keeping it organized is usually done in one of three ways. In the popular Ethernet system, pioneered by Xerox and now used by many companies, each device on the line simply "listens" for a clear signal before sending its package of data. If two packages accidentally collide, both devices wait and try again.

In token-passing systems, on the other hand, collisions are impossible because no device is allowed on the network unless it first takes possession of the single electronic "pass" that moves throughout the network. After a package of data reaches its destination, the pass is released for another device to use. Token-passing technology was introduced by Datapoint and is now widely used throughout the industry. In fact, IBM is expected to use token-passing technology in the long-awaited LANs it has developed for its larger systems.

Lastly, Star networks have central switchboards that perform much like their counterparts in a telephone network. Not surprisingly, Star networks were developed by AT&T.

MAJOR INDUSTRY TRENDS

It is unlikely that the trend toward general purpose LANs developed by independent suppliers will be threatened by the introduction of more comprehensive products by computer makers. The major computer firms are way behind the independents and have their hands full developing networks for all their own equipment. Rather than attempting to develop general purpose systems as well, the computer makers are going out of their way to cooperate with independent LAN suppliers in order to ensure that their products are included in multibrand environments. The situation that is emerging is one of vendor-specific networks from computer companies being tied together by general purpose LANs. This situation suits everyone's interests very well and will almost certainly continue.

The call for LAN standards, which was very strong a year or so ago, will continue, but with less strength. De facto standards have emerged naturally in the marketplace without having been imposed upon the industry. Because the surviving technologies tend to occupy separate niches where they meet specific needs, further culling is unlikely. Although each of the popular LAN technologies has limitations, it is probable that additional changes in standards will be evolutionary rather than revolutionary. As a prominent network engineer said recently, "In the office at least we can now hook almost anything up to anything else with what we have. Forcing standards onto the industry at this point would actually limit flexibility." I think the present market leaders are far more likely to make use of new technology than to be pushed aside by it and they are the best bets for investors.

The only products in sight with the potential to dampen LAN sales are the latest digital PBXs. As I discussed in the last section, modern PBX equipment provides many LAN functions, in addition to its other capabilities, and will continue to find ready markets. However, PBX systems do not handle data quickly enough for large-volume computer communications and will not replace LANs in most applications. Although the data handling capabilities of PBX systems will improve, the trend is clearly toward PBX to LAN connections rather than PBXs replacing LANs.

Several technological advances are becoming available this year that will make LANs increasingly popular. Complex electronic circuits with expanded functions are now being produced on single chips. As a result, special features that will link popular but formerly incompatible devices will reach the market for the first time this year. Lastly, fiber-optic LANs have moved out of the laboratory and are ready for many high-volume data transfer applications.

LAN COMPANIES AS INVESTMENTS

LANs are now available for many common computer and office automation systems. Sales are increasing and will accelerate as prices continue to drop and capabilities increase. Very powerful yet easily installed LANs will soon be sold off-the-shelf through normal distribution channels. At that time, LAN profits will soar.

General purpose LAN suppliers look particularly attractive at this time. Their market has been developing since the beginning of the computer era and is in no danger of being saturated with products for the foreseeable future.

LAN Companies

Company	Symbol	Exchange
3Com	COMS	OTC

Comments: A leading supplier of general purpose, Ethernet-based LANs for IBM, Texas Instruments, Hewlett-Packard, and other 16- and 32-bit personal computers and workstations. Competes with IBM's own PC LANs products but has an excellent chance to be successful because 3Com's networks permit many non-IBM peripherals to be included in the system.

Company	Symbol	Exchange
Network Systems	NSCO	OTC

Comments: Network Systems is the leading supplier of high-speed, general purpose LANs for the top end of the market. The need for high-performance LANs is growing very quickly due to the proliferation of powerful computers for scientific, factory automation, and other demanding applications. Excellent performance in 1984 and early 1985. Outlook is excellent.

Company	Symbol	Exchange
Sytek	SYTK*	OTC

Comments: An initial public offer of stock for this company is expected in late 1985. Sytek was selected by IBM to supply LANs for its personal computers. Contracts for additional products may follow. Well worth watching.

Company	Symbol	Exchange
Ungermann-Bass	UNGR	OTC

Comments: This company is a leading general purpose LAN supplier with products for every segment of the market. In October 1984, Ungermann-Bass accepted an offer from General Electric to jointly develop badly needed LANs for factory automation applications. Because GE is making major inroads into factory automation at every level, the joint venture has the potential to keep Ungermann in the chips for years. The potential for an outright buyout by GE appears substantial.

*Proposed symbol.

CHAPTER 15

The Market Expands for Data Communication Equipment

In previous sections we examined local area networks (LANs) and private branch exchanges (PBXs), which are revolutionizing voice and data communication within offices and factories. The widespread adoption of computers and computer-based devices has also created an enormous market for equipment that facilitates data communication over much greater distances, such as across a campus or from city to city.

Many sophisticated technologies exist for longer distance data communication, including satellite relay systems and microwave transmission, but most data communication occurs over standard telephone lines and other ground connections. For most hookups, modems, multiplexers, and other data communication devices are essential and are in great demand. As a result, the industry that supplies data communication equipment is growing rapidly.

MODEMS ARE ESSENTIAL

As mentioned in the section on PBXs (Chapter 13, this section), modems change the digital pulses produced by computers into tones that can be transmitted like speech over existing phone lines. A modem at the other end converts the tones back into pulses for the receiving machine. Without modems, data transmission could not occur except where special data links that permit the direct transmission of digital signals have been installed. Because such links are presently in short supply and are expected to remain so for the remainder of this decade, the modem industry should find itself with an assured and expanding market.

MULTIPLEXERS PAY FOR THEMSELVES

Multiplexers make it possible for several computers, terminals, and other digital devices to use the same line simultaneously, which reduces data communication costs substantially. Multiplexers are used by Regional Bell Operating Companies (RBOCs) and long-line carriers as well as individual companies with large data transmission needs.

The market for modems and multiplexers grew approximately 35 percent in 1984, a rate that will undoubtedly continue at least through 1986. By 1988, the market is expected to account for $10 to $12 billion of total telecommunication revenues. Although the technology may lack the glamor of satellites or laser links, the profits it generates should warm the hearts and pockets of the most sophisticated investors.

SUSTAINABLE GROWTH

Demand for specialized data communication equipment will certainly prove to be long lived, not only because of the improved economy, but also because technological and productivity trends greatly favor growth in this area. Lower prices and improved capabilities of computers and computer-based equipment are having an enormous effect on sales of such machines and consequently on the demand for all data communication equipment. The widespread adoption of LANs and PBXs makes the use of computer equipment even more cost effective, adding additional fuel to the fire.

In response to the unprecedented demand for data communication links, the new RBOCs, and the major long-line carriers including AT&T and MCI are adding data handling capabilities as rapidly as possible. These companies add considerably to the market for modems and multiplexers, both for their own use and for remarketing to their customers.

To see the dramatic effect that increased use of computers and computer-based machines is having on the market for data communcations equipment, one needs only to look at the explosive growth of data compatible T1 "telephone" circuits being installed by the telecommunication companies. In some sections of the country, the waiting list for a T1 hookup is well over a year. The market for T1 multiplexers alone is expected to grow 45 percent annually.

Personal and home computer users represent another huge market for data comunication equipment, especially modems. The growing popularity of low-cost networks with expanded services compounds the demand for such devices. Microcomputers in the office represent another growing market. Executives in the modem industry report that their products are the second most popular microcomputer accessory after printers. Last year, 530,000 units were sold. The number may reach 8.4 million by 1988, a compound annual growth rate of 73.7 percent.

Perhaps the most bullish factor for modem and multiplexer sales is their low cost and quick payback periods. A multimillion dollar computer system can be linked together over thousands of miles by equipment that might cost less than $50,000. Such low prices simply do not retard sales. Because the equipment allows many machines to use the same line, it often pays for itself in as little as three months. Even with high interest rates and the threat of a recession, modem and multiplexers are clearly cost effective and will continue to sell well.

Datacom Equipment Companies

Company	Symbol	Exchange
Cermetek Microelectronics	CRMK	OTC

Comments: Cermetek manufactures low-speed programmable modems for use in small computers. The company is a leader in the development of modem components for sale to original equipment manufacturers. Cermetek has had its ups and downs but it could profit from the popularity of microcomputer modems, which should become standard features in the new PCs scheduled to appear soon.

Company	Symbol	Exchange
Digital Communications Associates	DCAI	OTC

Comments: Digital Communications supplies multiplexers, network processors, and other network management devices. The company offers equipment for all levels of computers, from micros to mainframes. A specialty is the personal computer to mainframe links that have been very successful.

General Datacomm	GDC	NYSE

Comments: General Datacomm is a world leader in the data communications field. The company supplies modems, multiplexers, and fiber-optic equipment for transmitting data using every available link from telephone lines to satellites. The company sells high-capacity equipment to telephone companies and markets products to end users as well. The firm's new multiplexers are especially popular because they handle voice as well as data, making expensive duplication of equipment unnecessary. Foreign sales are substantial and should improve even more as the dollar declines in value.

Micom Systems	MICS	OTC

Comments: Micom supplies a wide range of data communication equipment primarily for micro- and minicomputers. The company publishes the well-known "Black Box" catalog of datacomm products which have been very successful. Products include modems, multiplexers, and data concentrators. Micom has a very practical approach to data communication that customers appreciate. Profits have been substantial.

Timeplex	TIX	NYSE

Comments: Timeplex produces modems and multiplexers. In addition, the company builds turnkey data communication networks to order.

CHAPTER 16

Satellite Business Communication Comes Down to Earth

The concept of artificial satellites was first proposed over 200 years ago by Isaac Newton. He pointed out that if an enormously powerful cannon could be placed above the earth's atmosphere, it would theoretically be possible to fire a cannonball around the world.

Newton's cannonball, of course, would start to fall toward the ground the moment it left the muzzle of the cannon. However, if its speed was just right, the curve of its fall would match the curve of the earth and it would become a satellite in orbit.

The speed needed for a satellite to achieve orbit above the earth depends on altitude. If the speed is too low for the height, the satellite will fall to the earth. If the speed is too great, the satellite will shoot off into space.

STATIONARY SATELLITES

At 22,500 miles above the earth's equator, the speed needed to keep a satellite in orbit exactly matches the speed at which the earth turns. By having the two speeds synchronized, the satellite stays over one spot on the earth as if it were a kite on a string.

Arthur C. Clarke, the great science fiction author, was the first to propose that "stationary" satellites be used to relay messages around the world as cost-effective alternatives to telephone and telegraph cables. With this idea the communications satellite industry was born. Today, of course, both human voices and computer data regularly travel from place to place on earth by first traveling through space.

THE SATELLITE DATA COMMUNICATION INDUSTRY

There are three segments to the satellite data communication market, distinguished primarily by the amount of information they distribute, the speed with which the information is transmitted, and the cost of the equipment.

At the high end of the market are the major telecommunications companies and large business concerns that move huge volumes of data at blinding speeds. This market is served mainly by the customers themselves or by Comsat (CQ, NYSE) through Satellite Business Systems, which serves many of the Fortune 100 companies.

Earth stations for the high-end market involve the use of large disk antennas and complex electronic systems. Because installation costs usually begin at $200,000, with proportionally high satellite user fees, this market is not expected to grow as quickly as many observers once expected. Indeed, Comsat has just announced its intention to sell its interest in Satellite Business Systems, which has not been profitable.

In the middle of the satellite business communication market are systems that typically cost $60,000 and up. The mid-sized market is much larger than that for high-end equipment and services, and is served mainly by American Satellite Company and Vitalink Corporation, both of which are privately held. Although the growth potential for the middle level of the market is considerably better than for the high end, it is well below the potential growth of low-end systems.

The bottom-end market is composed of companies that have

many offices dispersed over a wide area with whom daily communication is needed. In many cases the communications must travel only one way—from the home office to the branches, for example. Systems consist of a small dish about the size of a garbage can cover that is mounted on the roof or even pointed out the window in the direction of the satellite. The dish is connected to an electronic decoder that resembles a home stereo receiver, which in turn is connected to a microcomputer. Total costs can be under $10,000.

The cost-effective nature of small-scale satellite business systems is substantial. Expensive, leased telephone lines to all receiving stations are eliminated. Because the satellite spreads its signals over the entire nation, there need be no additional service charges for extra stations that simply receive information. Of course, there are no long distance charges.

INVESTMENT OUTLOOK

Until recently, the low-end satellite business communication market existed only on paper. There simply wasn't a way to supply equipment and services at moderate costs. However, marketing specialists predicted that the communications satellite company that found a way to crack the price barrier would find a large and ready supply of customers. They were right.

The low-end satellite business communication industry currently has only one pureplay available for investors, Equatorial Communications. This firm has both one-way and two-way systems available and has been quite successful. Additional firms will certainly enter this expanding area in the coming months and should be investigated by investors.

Although the investment potential of the low-cost satellite business communication field is substantial, there is a cloud on the horizon of which you should be aware. One day, perhaps five to seven years from now, inexpensive fiber-optic ground links will become cost competitive with satellite communication systems. As such links come on-line, the market for satcom systems will shrink. Meanwhile, this booming field will generate enormous profits.

Small-Scale Satellite Communication Companies

Company	Symbol	Exchange
Equatorial Communications	EQUA	OTC

Comments: Equatorial is the leading supplier of low-end satellite business communication systems. The company is selling its one-way systems to an eager market and should soon be ready with its two-way equipment. Incredibly, prices for one- and two-way systems are well under $4,000 and $8,000 respectively, and are expected to drop once high-volume production begins. Very good potential for appreciation.

Company	Symbol	Exchange
M/A-COM	MAI	NYSE

Comments: M/A-COM is a broader satellite play than Equatorial. The company has low-end systems ready to market but is also involved with other satellite and data communication products. In addition, the company is involved in television broadcasting, cable TV, and defense communication equipment. Established in 1950 as Microwave Associates, M/A-COM has established a good track record.

Section 4

A New Generation of Computers Creates Attractive Opportunities

CHAPTER 17

A Transformation Is Occurring in the Computer Industry

Has anyone noticed that in recent months, we have frequently seen conflicting reports about the computer industry? On one hand are figures indicating that sales among the leading makers are not what they used to be. On the other hand, user surveys indicate that the number of computers being installed is up sharply. If both sets of data are correct, then computer users have been buying their machines from companies other than the leading makers.

Both studies are correct, and they indicate that important changes are occurring in the computer industry. The new computers that are appearing in the marketplace differ in so many important ways from old-generation machines that they are rapidly pushing the old computers aside. The basic structure of the industry will certainly change to reflect the increased status of the new machines and the companies that produce them.

A few years from now, many of the computer industry changes that are now in their early stages will be apparent. Before that time, investors who take the trouble to become familiar with the opportunities these changes present will be in an ideal position to make careful selections for substantial long-term growth.

OUR MAINSTAY COMPUTERS JUST CAN'T COPE

The great majority of the computers in use today have been designed to handle any type of work likely to be encountered in 98 percent of all business and industrial applications. Whether the machine is a personal computer, a large mainframe, or any size in between, the chances are if it is over a year old, it can do word processing, database management, and complex mathematics with equal facility. Unfortunately, it may not do any particular task well enough.

The problem is that the various tasks that computers are called upon to do are vastly different from each other and need to be handled in entirely different ways. In order to make it possible for a single computer to do everything, many compromises must be made in regard to overall efficiency, speed, and power. It is simply not possible to optimize the handling of any particular function in a general purpose machine.

Until quite recently, the capabilities of general purpose machines expanded faster than the growing needs of the market. As the technology matured, however, growth slowed. Many businesses and industries began to find themselves with needs that surpassed the capabilities of even the most powerful machines. Complex databases were developed but couldn't be managed efficiently. Modern engineering problems in aerospace, semiconductor design, and other sophisticated fields could take days to process. Progress in many critical areas ground to a halt. All too often, long-awaited computers were met with groans of disappointment when they were finally released.

The need for better machines went largely unnoticed as it is difficult to measure something that doesn't happen, such as a sale that doesn't occur. The latter situation is all too common with top-of-the-line general purpose computers because they can't function well in the newest and fastest growing markets. In any event, the drop in sales by the mainstay computer firms was completely misinterpreted by many observers, who took it as a sign that the computer market as a whole was reaching saturation.

BIG CHANGES ARE STARTING TO OCCUR

At the same time general purpose machines were reaching their peak, new chips, new techniques for using those chips, and vastly lowered prices all came together to create a variety of new and unbelievably powerful machines that meet the needs of the most demanding applications. The latest computers differ markedly from their general purpose predecessors in that they tend to be optimized for performing either numerical or database functions. In many cases specialization goes one step further, yielding, for example, machines that will do one particular class of math problems many times faster than anything else on the market.

It is now becoming cost effective for a company, even a small company, to purchase a state-of-the-art computer to serve its primary needs and a general purpose machine to do everything else. As we will see, the new technologies are now being applied to general purpose computers as well, which will revitalize that industry by the end of the decade.

Not surprisingly, the demand for more powerful yet affordable computers has become enormous, and the new machines are being snapped up by computer-starved organizations in every industry. Many of the young companies that make the modern computers are running at full capacity and are growing well over 100 percent a year. Some of the most successful firms have never spent so much as a dime on advertising and yet they can't keep up with demand. Many others that aggressively demonstrated their wares have presold every machine they can possibly make for months to come.

Far from being saturated, the computer market is in the process of expanding. However, the greatest demand is rapidly shifting from traditional machines to a new generation of computers that are well worth examining in detail.

FRONT-END MACHINES AND SPS SOFTWARE ALSO BOOM

In addition to the growth occurring in the market for new computers, related opportunities are opening up with modern automated data entry (ADE) devices. ADE machines break the front-end bottleneck in data processing by making it possible to quickly enter text and images into a computer without going through the keyboard. ADE devices may well be the hottest development in computer peripherals since printers, and will be worth millions of dollars within a few years.

Lastly, there are now ways to squeeze more efficiency out of a computer. System productivity software (SPS) is now available that does for a computer what an executive time management program does for a person, and more. This exciting class of software is just now becoming well known and promises to be a major growth area for the next several years.

CHAPTER 18

New Computers, New Investments

There are many new computers on the market from companies that hold great promise for investors. During the next few years, the new developments discussed here will make millions of dollars for their backers. Some of the areas represent totally new technology. Others involve older ideas that have been vastly upgraded for modern needs. In both cases, profits should be abundant.

DATABASE MACHINES

Almost all large corporations suffer from information overload. This problem has grown beyond the capabilities of general purpose computers, however large. Insurance companies, credit organizations, government agencies, and others with huge quantities of data to process have been especially hard hit by the limitations of the present line of mainframe computers. For many firms, the problem has become so severe that it is having an impact on profitability.

The traditional cure for undercapacity is to install new software. Although better software can postpone the problem, if the company continues to grow, its increased database needs will eventually tax the system.

Oddly enough, in many situations, adding more computers to the system will not adequately cure undercapacity. When more computers are purchased, databases must often be split, which makes working with them very inefficient. Of course, the machines may be connected to permit the sharing of information, but communication bottlenecks and other problems can render the system equally inefficient. As one insurance executive explained to me recently, "Look, it's like having a motorcycle that isn't powerful enough. You can soup it up but if that doesn't do the trick the only solution is to get a bigger machine. Owning two won't solve your problem."

New hardware solutions to problems of information overload are now available. Specialized database machines are revolutionizing information management. Several of the new computers contain multiple processors that break big jobs into small parts that can be handled simultaneously, much as a team of secretaries, each with a file drawer to search, can quickly find any needed information.

Database machines are designed to work with general purpose computers by relieving them of almost all information management functions. Since they are streamlined for just one task, they are much less expensive than any other hardware solution to undercapacity. As a bonus, general purpose machines run more efficiently and become more useful once they are relieved of overtaxing database management responsibilities. Customers find database machines make a lot of sense from every perspective.

Database management machines are selling rapidly to businesses with information overload problems. Because they are new, word of their cost-effective capabilities hasn't spread widely. As a result, I expect that their period of greatest growth still lies ahead. Although there is always the chance that a large computer company will jump into the market, that market is expanding so rapidly that there should be business enough for the major players for many years.

Database Machine Company

Company	Symbol	Exchange
Britton-Lee	BLII	OTC

Comments: Britton-Lee is the leading pureplay in database machines. The company's first products were initially offered in early 1981 and are now becoming well known and firmly established. The company has done well since its 1984 initial public offering (IPO).

NEW FAULT-TOLERANT SYSTEMS

In many applications such as on-line bank transactions, travel reservations, and industrial automation, a computer failure would be devastating. As a result, special computers were developed several years ago that could tolerate faults and still function. Most such computers were expensive and difficult to program, but they did offer the necessary reliability.

About three years ago, however, the market for fault-tolerant computers ran up against the same limitations with existing designs as did owners of general purpose computers. Customers began demanding more power, more speed, and less expense from their systems. As a result, industry leaders and a host of start-ups undertook crash programs to bring fault-tolerant computers up to date. The result is a revitalized industry that is finding itself with the profitable task of replacing much of the existing base of machines. In addition, the new systems are finding totally new markets where affordable fault tolerance is seen as a highly desirable extra feature on a first-class computer.

Fortunately for investors, the race hasn't been easy. Unlike other computer markets in which the industry giants will probably make an appearance, it looks as if the big boys have thrown in the towel regarding true fault tolerance in a top-performing computer. IBM's decision to buy its fault-tolerant systems from Stratus was the first time the company has gone outside for an entire line of computers. DEC, Hewlett-Packard, and other top firms have announced systems only for second-tier applications where high availability rather than true fault tolerance is needed.

Fault-tolerant computers achieve fail-safe reliability by using modern chips containing redundant circuits, alternative data paths within the machine, and often multiple processors. However, some systems rely almost exclusively on hardware to achieve fault tolerance and others place much of the responsibility for total reliability on software checking systems. Generally, hardware-based systems are more expensive than software-based systems but are easier to program. On the other hand, software-based systems, which rely more heavily on elaborate error-checking procedures than on hardware solutions, are less expensive and are far easier to expand. The choice of which system to install depends on the application. The market appears to be split about evenly between the two types, although hardware-based systems should take the lead within a year.

It appears from looking at market statistics and the unprecedented action of IBM that the demand for the new fault-tolerant computers is expanding rapidly and will continue to do so for several years. The field represents an attractive area for long-term investors.

Fault-Tolerant Computer Companies

Company	Symbol	Exchange
Stratus Computer	STRA	OTC

Comments: Stratus is a top maker of fault-tolerant computers. The company uses a hardware approach to achieve system reliability, which facilitates the development of software. Stratus was selected by IBM to supply products for resale—the first time Big Blue has gone outside for complete computer systems. Growth has been excellent.

Company	Symbol	Exchange
Tandem Computers	TNDM	OTC

Comments: Tandem is the leading maker of fault-tolerant computer systems for on-line applications. The company uses a software approach to total reliability. Although such systems require more care in the development of application software than do hardware-based systems, Tandem supplies excellent software development tools that simplify the process. In only nine years, Tandem has grown into a half billion dollar company and is now a member of the Fortune 500. A new-generation computer is proving to be very successful.

TECHNICAL WORKSTATIONS

As the situation with database machines illustrated, often the most cost-effective way to improve the capabilities of a general purpose computer is to down-load difficult tasks to devices specifically designed to handle them. Nowhere has this technique found wider acceptance than with specialized workstations for CAD/CAM, CAE, computer graphics, and several scientific applications. Although the new workstations have not escaped the attention of Wall Street, we feel their long-range potential is much greater than is generally recognized. Indeed, the workstation market, which reached $500 million last year, will very likely reach $3.2 billion or more in 1988.

Although technical workstations are designed to be part of a larger system, they contain their own powerful microprocessors and are frequently referred to as supermicrocomputers. Such systems, with their advanced circuits and innovative designs, offer the processing power and software capabilities of minicomputers costing much more. Because they are designed to support multiple users and multiple application tasks, they are ideally suited for large, complex projects involving teams of specialists.

There is little doubt that supermicrocomputers are the ideal machines for the majority of today's fastest growing technology industries. As prices continue to fall and the number of applications increase, their market will expand to include many other areas presently served by more expensive small and medium-sized minicomputers. Technical workstations are out of the starting blocks, but it is apparent they have yet to reach their stride. The industry should continue its rapid growth for several years, although short-term results are likely to be cyclical.

An important step toward capturing the huge minicomputer industry is just getting underway. Most of the major supermicro suppliers now offer machines with standard operating systems, including UNIX. Although the lack of an industry standard wasn't a limiting factor in high tech applications, it did isolate the machines from mainstream corporate America.

Now those barriers are coming down, which should have predictable results.

One caution is in order. Of the various new-generation computers discussed in this section, supermicrocomputers are the most likely to attract competition from the established computer makers. While it is true that IBM ran into a brick wall trying to come up with a first-class fault-tolerant system, and the number of people who can create a supercomputer could fit into a bus with room left over, for all their sophistication, supermicrocomputers can be duplicated. The besieged minicomputer makers are especially likely to do so in an attempt to recoup market share lost to workstation companies. However, they will find themselves trying to play catch-up with an accelerating target.

Technical Workstation Companies

Company	Symbol	Exchange
Apollo Computer	APCI	OTC

Comments: Apollo is the leading supplier of technical workstations. The company's machines consist of 32-bit computers that are tied together by an excellent high-speed network, which has the effect of creating a super system out of the individual parts. Apollo's success has attracted many competitors and will undoubtedly attract many more in the future. Although the company will need to roll up its sleeves and fight harder for market share, it has considerable momentum and will be difficult to dislodge.

Company	Symbol	Exchange
Encore Computer	ENCC	OTC

Comments: Encore is aiming its efforts primarily at scientific computers. However, the company intends to market high-speed workstations as well, which should find a ready market. (See the additional Encore entry later in this section.)

ARTIFICIAL INTELLIGENCE COMPUTERS

Of all the new-generation computers, AI machines are the most unique and have the greatest potential for long-term growth. However, they are also the most likely to evolve substantially from their present form as the young technology progresses toward true fifth-generation machines. But AI holds considerable promise for knowledgeable investors who take positions wisely and follow developments in the field. Indeed, investors who compound a few good decisions in this area over the next several years should make a very substantial amount of money.

Unfortunately, few investors understand artificial intelligence or the computers that support the process, which makes wise investment decisions very difficult. Most often AI machines are confused with supercomputers and scientific computers, but in truth they are vastly different. To see the difference, it may be useful to look, not at artificial intelligence, but at actual human intelligence.

When people solve math problems or look up data in a reference book, they are performing functions that are common to present day computers. With such tasks, both machines and people achieve results in much the same way. A limited number of specific rules are followed and work progresses sequentially from start to finish. Of equal importance, answers are specific. At no time is it necessary to guess or modify the rules.

When people depart from cut-and-dried thinking and take on tasks for which there are no well-defined procedures, thinking processes change dramatically. People usually begin solving the problem by starting with a likely solution path that will change frequently as work progresses. In addition, the information that must be considered isn't usually arranged in order—it is located by association, not by sequential thought. As a result, thinking is accompanied by a stream of ideas and images whose significance is evaluated as it flows by. Inferences are made and answers are chosen because they are the best choice, rather than an absolute conclusion. For such thinking, standard computers, including the most powerful models, are poorly suited.

Solving AI problems efficiently requires machines that can approximate the way humans deal with complex tasks. The first step toward creating such machines was to develop a language that made it possible to use inferences and deal with probabilities rather than absolute values. The second step was to develop hardware with chips and circuits that could make many more memory references than other computers and could do so using selection procedures based on relationships rather than on specific addresses. Both developments are now a reality, and the machines that contain the new technologies are most impressive.

For the next several years the major thrust in artificial intelligence will be toward creating expert systems in many fields. Such systems will duplicate the ways humans solve very specific but complicated problems. Projects to design electronic experts in areas as diverse as medical diagnosis, divorce law, semiconductor design, and missile guidance are already underway and have reached unbelievable levels of sophistication.

Within a very short time, certainly by 1988, expert systems will be found in a growing number of applications. In the past year, several AI software tools have been introduced that make it possible for customers in many industries to create expert systems. Significantly, crash programs are underway to use the new software tools to create high-powered expert systems that will themselves be used to create other expert systems. As one spokesman for an AI firm explained to me recently, "It should be possible within a few years for a human expert in almost any field to teach a computer to perform his job." He went on to say that expert systems would be to many office personnel what robots are rapidly becoming to blue-collar workers. But take heart. If you play this field properly, you'll retire long before your silicon replacement arrives.

AI Company

Company	Symbol	Exchange
Symbolics	SMBX	OTC

Comments: Symbolics is a pureplay in artificial intelligence computers. The company appears to be the leader in the field, with well over 50 percent of the market. The company's systems work well with machines made by IBM, Digital Equipment, and other companies, which gives Symbolics a big advantage in most markets. In addition, Symbolic's machines will run standard science and math languages, which makes them multipurpose computers whose cost most firms can justify. The company's future appears excellent, but please remember that the AI field is still young and may experience many changes as it matures.

SUPERCOMPUTERS

Few types of computers have garnered as much attention as supercomputers, and for sound reasons. Supercomputers process numbers at blinding speeds and have become essential in many areas of scientific research. Supercomputers are also needed for many vital technologies including aircraft design, semiconductor engineering, drug and genetic modeling, and nuclear weapon simulation. Indeed, not only can the presence or absence of adequate supercomputers make or break important individual projects, but supercomputers are also beginning to play a crucial role in maintaining our national security and economic health. At the local level they can have a very visible impact, as the prosperous residents of Seattle learned when a new supercomputer made it possible for Boeing to create its popular 767, and thousands of jobs.

Unfortunately, meeting our needs for supercomputers isn't as easy as simply building more machines. Although it may come as a surprise to some readers, our present generation of supercomputers is based on technology that is almost 10 years old and is operating very close to its maximum level of performance. As a result, supercomputer users are beginning to find themselves with the same problems of undercapacity that other organizations have with their out-of-date machines.

However, new technologies are now becoming available that will be capable of handling our greatly increased supercomputer demands. Reviewing the new machines and their investment potential should be quite valuable for those who wish to share in the substantial growth occurring in the supercomputer industry.

In order to better understand how old- and new-generation supercomputers work, let us start with a simple analogy. If a contractor needed to dig a hole big enough for a septic tank, the simplest way to get the job done might be to give it to a skilled man with a shovel. If the work didn't go quickly enough, the job could be given to a more powerful man. If that didn't do the trick, better tools could be brought in for the different types of soil encountered. However, if the most powerful man working with the best tools still wasn't fast enough, there would only be one solution: putting two men on the job.

However, there are many ways to use two men. One man could be put in the hole to dig while the other remained outside to hand in tools like a surgical nurse and keep the soil cleared away. Alternately, both men could be placed in the hole. One worker could use a pick to loosen soil and the other a shovel to toss it out of the hole. The problem with this method is that only one man is actually digging the hole. If both men could share the job of shoveling but work independently in different areas, the output would be increased, providing the soil was easily shoveled. Of course, good coordination would be necessary with any of the multiworker options or the project might look like a Keystone Cops movie.

Let us carry the situation one step further to make an important point. The contractor might be tempted to place more than two men on the project to get it done even faster. However, the more men that are involved, the harder it becomes to coordinate them, and the more likely it is that the job will actually take more, rather than less, time to do. In any event, there are just so many people that can fit in a hole. In other words, there are just so many ways a job can be broken down into individual parts.

Referring again to supercomputers, the old-generation machines use single processors. These processors are unbelievably powerful in comparison to the ones found in lesser machines, largely because their circuit elements are packed together more tightly, which permits them to exchange information at a higher speed. In addition, the processors are accompanied by the latest tools and techniques for handling complex problems. Equations are rarely handled step by step, as we learned to do them. Instead, they are broken down into similar tasks and sent through specialized "pipelines" so that many calculations may be done at once. The system feeds numbers and commands into the processor using multiple circuits, which keeps information moving as fast as it can be used.

Supercomputer speeds are very fast indeed—as many as 100 million scientific calculations a second for problems that can be organized to make use of a supercomputer's unique operating style. However, single-processor machines have evolved to an advanced level of sophistication and are unlikely to be made significantly faster. Major advances are especially unlikely now that IBM, and almost everyone else, has given up trying to develop the last great hope, the super fast Josephson-junction computer chip. For all practical purposes, the present technology is fully mature.

As was the case with the contractor whose job was too big for one man, manufacturers of high-speed computers are turning to multiple approaches but are having many of the same types of problems with organization and coordination. At the present time, the few top multiprocessor systems coming on the market divide a project into separate parts that are processed concurrently or, as it is sometimes termed, "in parallel." Most such systems use two to four processors. Each processor is very powerful and is organized internally like a single processor computer. These new tightly coupled multiprocessor systems will be the heart and soul of the high-speed computing industry for the next few years and are the ones investors should watch most closely.

Less powerful multiprocessor systems work like the two men in the hole, when one had a pick and the other a shovel. In both cases the *type* of work is divided, not the actual project. In some cases, having different microprocessors do different types of work can boost efficiency. However, in most cases, throughput is not increased nearly as much as would occur if each of the processors would take part of the project itself and work on it concurrently with the other processors.

Some "massively parallel" computers will be out soon, which will have dozens, perhaps thousands, of processors. Because each processor will have less to do than would be true in a smaller system, each one needn't be particularly powerful. As a result, massively parallel computers are expected to be quite inexpensive by supercomputer standards.

Whether massively parallel systems will be useful, however, is another matter. As with the problem of having too many men trying to dig a hole, for most applications, using massively parallel systems will almost certainly be gross overkill. Not only are few jobs complex enough to be subdivided into more than 32 (or at the most 64) parts, but writing software to coordinate the various elements will be a horror. For some highly complex tasks such as weather calculations, that are performed the same way time after time, massively parallel systems will probably be useful and worth programming. Unless special artificial intelligence programs can be developed to write the software for these machines, however, they are unlikely to become the mainstay of supercomputing.

The projected growth in supercomputer sales is impressive. In early 1985, the market was worth approximately $300 million a year, but it is expected to grow to $1.5 billion by 1990. If one expands the definition of supercomputing a bit to include the machines discussed next, the growth will be truly staggering.

Supercomputer Companies

Company	Symbol	Exchange
Amdahl	AMH	ASE

Comments: Amdahl is the leading manufacturer of IBM-compatible mainframe computers. The company is 49 percent owned by Fujitsu and will market that firm's supercomputers in the United States (see Fujitsu note, below). The extent to which supercomputers will contribute to revenues is unknown, but is likely to be modest at first.

Company	Symbol	Exchange
Cray Research	CYR	NYSE

Comments: Cray is the world's leading supplier of supercomputers and has every chance of remaining so. The company's success with the single processor, Cray-1, has been phenomenal. New products with multiprocessing capabilities, including the X-MP and the Cray-2, should ensure this company's continued success.

Company	Symbol	Exchange
Denelcor	DENL	OTC

Comments: Denelcor is America's third largest maker of supercomputers after Control Data (through its subsidiary, ETA Systems) and Cray Research. The company has been spending large sums on R&D and has been running in the red. Because of its losses, Denelcor receives much less attention than its larger competitors. This is a long shot but is very definitely a company to review prior to making an investment in this area.

Fujitsu

Comments: Fujitsu is Japan's leading computer maker. Systems range from one-chip microcomputers to mainframes, and now supercomputers. The company will market its new supercomputers in the United States through Amdahl Corp. The Japanese computer giant has yet to make much of a dent in the supercomputer market, but it hasn't been at it very long. It should be noted, however, that Fujitsu's other computers have not been big sellers in the United States.

Control Data (CDA, NYSE) is a leading supplier of supercomputers via its subsidiary, ETA Systems. Unfortunately, Control Data is quite diversified and cannot be expected to appreciate substantially from its supercomputer sales.

ARRAY PROCESSORS AND SCIENTIFIC COMPUTERS

Supercomputers are expensive. The least costly model will usually exceed $5 million, once it is fitted out and ready to run. Top-of-the-line systems will easily soar past $15 million. As may be expected, as many as three fourths of the organizations that need supercomputers can't afford them. Those organizations represent a huge and frustrated market, ready to snap up the first effective high-speed computers they can find.

Fortunately, machines intended for the low-end supercomputer market have recently been developed and are beginning to come off the line. Newly developed scientific computers and the latest generation of array processors both contain many of the internal features common to supercomputers, including state-of-the-art chips, pipelining, multiple circuits, and even multiprocessing.

The key to the exploding success of the new high-speed machines is their cost effectiveness. The best of the new devices will operate at about one half the speed of a supercomputer but will cost only a 10th as much. As a result, the cost per calculation is far less with the new machines than with supercomputers. For many thousands of organizations with substantial computational requirements, near-supercomputers make a lot of sense. "We don't care if our new scientific computer takes three hours rather than one and a half hours to do one of our engineering studies. We're happy we can do them at all," said one user whose opinion characterized the general sentiment of the market.

Modern array processors are the least expensive high-speed number crunchers available. The capabilities of the latest machines are vastly improved over earlier models and often extend well into the supercomputer range. Not surprisingly, the new array processors are finding ready markets wherever they can be used. However, the increased potential of these new machines has gone largely unnoticed outside the computer industry.

Array processors are designed to function as supercalculators and nothing else. Because they lack a computer's ability to swap files, develop programs, and interact with users, they must be connected to general purpose machines. In addition, array processors are best suited for highly repetitive problems such as those found in medical imaging, real-time simulation, engineering analysis, and geophysical prospecting.

Array processors can't do everything, but an increasing number of organizations are finding them ideally suited to their needs. The expanding demands are causing the market for array processors to grow a healthy 30 percent to 35 percent a year.

Scientific computers are much more versatile machines than array processors and are best thought of as small-scale supercomputers. Unlike their big brothers, however, scientific computers use inexpensive off-the-shelf components. In addition, scientific computers are much easier for rank and file scientists and engineers to operate because they use standard programming languages and are therefore able to run most existing application software.

In terms of both price and performance, scientific computers fall between top-of-the-line minicomputers and supercomputers. Until scientific computers were introduced this past year, the computer gap was a huge no-man's-land for anyone that couldn't make use of array processors. The number of computation-intensive organizations has grown by leaps and bounds and represents a very large market for the new machines.

The demand for scientific computers is so large that we must not speak of the market as developing, as we would with any other new technology. The market for scientific computers is already developed, it is huge, and it knows what it wants. It will rapidly absorb every high-quality, cost-effective machine the manufacturers are able to produce for a long time. Although dollar estimates for the annual value of the market vary widely, I feel comfortable in saying that it should be at least 5 times the size of the market for supercomputers, and I won't raise my eyebrows if, a few years down the road, I find that 10 times would have been more nearly correct.

A-P and Scientific Computer Companies

Company	Symbol	Exchange
Analogic	ALOG	OTC

Comments: Analogic is a leading supplier of array processors and other data gathering and signal-processing devices. Not a pureplay, but the company will benefit substantially from the increased demand for array processors and related equipment.

Company	Symbol	Exchange
CSP (CSPI)	CSPI	OTC

Comments: CSPI is a leading supplier of array processors. The company offers 32-bit and very powerful 64-bit products primarily for imaging, molecular modeling, seismic, and CAE applications. CSPI has been quite successful.

Company	Symbol	Exchange
Floating Point Systems	FLP	NYSE

Comments: Floating Point is a leading supplier of array processors including new-generation models with speeds comparable to supercomputers. The company also makes two excellent scientific computers that are selling well. Floating point has a first-rate reputation for top-quality products and service and should continue to prosper. The company has a broader participation in high-speed computing than most firms in this industry.

Two additional scientific computer firms with excellent potential are discussed later in this chapter under general purpose computers.

RISC COMES OUT OF THE LAB

Also moving into the computer gap, at least at the bottom end, are reduced instruction set computers or RISC. RISC has been under development since the 1970s, primarily by IBM, and represents a change in computational strategy rather than a totally new technology. RISC products offer improved speed at affordable prices and may be successful. We'll need to wait and see.

RISC represents a return to basics. Early computers were so underpowered and short of memory that they were made to perform a bare minimum of functions. Complex tasks were accomplished by using a combination of the basic operations in much the same way we might use a simple four-function calculator to solve a complex algebra problem.

As computer science progressed, many additional functions

were added to the basic instruction set. However, the added features made the machines more expensive and, more importantly, the extra baggage kept the machines from running as quickly as would otherwise have been possible. But in the heyday of the present generation of machines, power was expanding so rapidly that no one cared if computers weren't running at maximum efficiency. The technology has since reached its peak. Consequently, in order to increase computational speeds, the extra baggage is being thrown overboard.

But RISC may be risky both for customers and investors. Quite a few computer experts fear that adapting the mountain of existing software to the new machines will be more trouble than the extra speed is worth. Of course, the companies that are developing the new computers downplay the software compatibility problem. They promise special translation circuits and software development tools to streamline the conversion to RISC.

Since the jury is still out regarding the potential of RISC, I urge readers to take a wait-and-see strategy. There is so much money to be made by being second in line for any new technology investment that there is no need to accept the hazards of being too early and possibly wrong.

RISC Company

Company	Symbol	Exchange
Hewlett-Packard	HWP	NYSE

Comments: Although IBM has done much work on RISC technology, Hewlett-Packard is doing more than any other public company to bring RISC to market. In fact, the firm scrapped an existing computer project in order to concentrate on RISC. The first product should be out in late 1985. Like all HP equipment, it will certainly perform well, but whether the machines will sell is another matter.

A NEW BREED OF GENERAL PURPOSE COMPUTERS

Computers are so complex that at first, every new development is applied very narrowly. Later, when the technology matures and becomes more widely understood, it can be applied to a broader range of applications.

It is now apparent that the latest and most promising of the new computer technologies is multiprocessing. At the present time, computer experts have their hands full developing the means to achieve true concurrent computing in mathematical applications. However, multiprocessing holds great promise for revitalizing general purpose computers and will certainly be applied successfully in that area before the end of the decade. In fact, a multiprocessor computer is now available that is organized to permit its use in every area of computing. Of course, very little of the software necessary to run it has yet been written, and the machine itself will probably undergo several revisions as the technology sorts itself out.

Whatever final form the new general purpose computers take, they will find the biggest market of all. When perfected, modern general purpose computers will return billions of dollars to investors who take the trouble to follow developments in the industry and who take positions accordingly.

New Generation GP Computer Companies

Company	Symbol	Exchange
Encore Computer	ENCC	OTC

Comments: Encore's public offering occurred in the spring of 1985. In late 1985 or in early 1986 the company expects to begin marketing high-speed computers and technical workstations, many of which will contain multiple processors. Encore's machines will be used primarily in scientific applications but appear to be adaptable to general purpose computing. The company's founders and chief officers are top computer people who are unusually qualified to make Encore a success. Definitely a company to watch.

Company	Symbol	Exchange
Flexible Computer	FLXXA	OTC

Comments: Flexible has been remarkably successful with multiprocessing computers. The company's systems are usually used in scientific processing but can be placed in general purpose situations as well. This firm has potential and should be followed.

Other public companies with substantial involvements in the new computer industry include: Star Technologies (STRR, OTC, scientific computers) and MASSCOMP (MSCP, OTC, scientific computers and workstations).

CHAPTER 19

New Software Boosts Computer Efficiency: Its Suppliers Reap the Rewards

In the last section I mentioned that for rapidly growing companies, improved software could only postpone the purchase of a new computer. However, in less demanding situations where computer undercapacity is a problem, better software can often provide a solution that will work for several years, if not permanently, and in the process save the firm well over a million dollars in new equipment. Fortunately, new software packages are now becoming available that greatly improve computer efficiency, and they are selling very well.

It isn't difficult to see why demand is growing rapidly for products that cure system undercapacity. This slowly developing condition can become crippling if left unchecked. Undercapacity can creep into any business, even if its growth isn't substantial, because greater computer awareness among staff members increases computer use. Labor-saving peripheral devices and personal computers are usually added one by one, until they also tax the system, and office efficiency drops. Expensive personnel wait longer and longer for the computer to respond to instructions. To make matters worse, people often adjust by not using the system for anything except critical projects. As a result, much work that might have been of value isn't even attempted.

The usual solution to undercapacity is to purchase a bigger computer or add another to the system. As we saw earlier, if the organization's business is heavily oriented toward database management or technology, it may solve its problems by adding specialized devices to the main computer. Whatever hardware solution is selected, though, it will be very expensive.

Purchasing more computer power was once the only solution to undercapacity, but that is no longer true. A few years ago, a number of computer specialists found that most machines ran well below their design capabilities because the system software that came from the factory did not use the machine's power to its best advantage. They found that computers, like people, could vastly increase their output if they were better organized and worked more efficiently. With that understanding, a new industry was born to create system productivity software, SPS. That industry has quietly matured and is now a multimillion dollar business.

IMPROVING COMPUTER EFFICIENCY

Much system productivity software is loaded directly into the computer's operating system. Except for minor procedural changes, people who use the computer are often unaware of the new software that is substantially improving their computer's performance.

One of the newest and most powerful of the SPS packages optimizes the running of other software programs. Some SPS packages interact with the user's business application software and make changes that allow the instructions to be processed faster.

SPS performance management packages schedule incoming jobs and user requests so that computers will run at maximum efficiency. Instead of the usual procedure of automatically taking jobs in the order that they are received, work may be scheduled to reduce turnaround time for everyone. Similar jobs are often merged and run together. Although in practice, the first come, first served rule is abandoned, the effect is to make each user feel his request has received the system's immediate and total attention. The real benefit, of course, isn't just the fast turnaround time for the computer: It is the increased productivity of the staff as employees spend more time working and less time waiting for the system to respond.

Several SPS packages will generate reports on the status of the computer system and how it is used. Often the reports will suggest methods by which the staff can improve operations. In many cases, small changes may have a big effect on efficiency. Most packages will list times during the day when capacity is strained and times when it is underused. Schedules can then be adjusted to balance the load.

In addition to scheduling jobs, SPS programs can coordinate the use of equipment to improve efficiency. One package is especially popular with organizations that have two computers because it automatically links them for large projects. When people have smaller projects to do, the systems are separated in order to double the effective capacity. Most SPS packages also optimize the use of printers and other peripherals.

A particularly useful class of SPS products speeds the flow of data within and between systems. The new packages usually make it unnecessary to duplicate data. Besides improving the performance of the system, the software greatly reduces the need for expensive storage devices. With all the data stored in one place, the most current information is available to everyone in the same form.

Having an organization's data in one place also makes it more secure. In fact, security is improved overall with new system productivity software that contains up-to-date safeguards. Incredible as it may seem, most of the system software supplied by hardware companies still comes with woefully inadequate security functions. Unfortunately, add-on security packages that don't become part of the system are often not much better. To be most effective, security functions must be built into the system software itself.

Of all the SPS capabilities, perhaps none is more valuable to the day-to-day running of a business than cost analysis. Special programs are available that track each job and report what it really costs in terms of both computer and operator time. Such information lets a company determine if a job is worth doing or if it is priced correctly. Many customers report that acquiring good cost analysis is like being handed a light in a dark cavern—they can see both the beauty and the pitfalls at once.

Some SPS products contain capacity management programs that help plan additions or changes to the computer system. The best of such programs simulate hypothetical computer systems, then analyze their efficiency and capabilities. Although hardware suppliers have similar packages available, they are sales tools and are not very popular due to their bias towards specific brands of equipment.

INDUSTRY AND INVESTMENT OUTLOOK

System productivity software has a much broader market than application software because its functions are more universal. No matter what an organization's business may be, SPS packages make its computer system work more efficiently.

The market for SPS packages is geographically broad as well. Unlike application software, whose instructions must be translated into the language of the user's country, English is the standard for system software. SPS packages developed in the United States may be sold, as is, around the world.

Because SPS is highly cost effective, the industry may be largely insulated from the effects of adverse economic conditions. Indeed, during bad times, system productivity software is especially appealing for those who need additional computer capacity since the alternative, to buy more hardware, is much more expensive.

The market for SPS is far from saturated. Less than 10 percent of the organizations with mainframe computers have first-rate system software installed.

System productivity software is often the most cost-efficient computer product a company can purchase. Not only is it priced well below a new computer system, but many of its features are not available from hardware manufacturers. Because it uses the customer's existing equipment, the organization's personnel do not need to learn a totally new system or substantially change their way of doing business.

It is little wonder that wherever overcapacity problems may

be solved with system productivity software, a sale almost always occurs. This trend will accelerate as more and more data processing managers learn about SPS and ask to try it before spending a king's ransom on additional computer equipment.

The long-term outlook for the SPS industry is very favorable.

SPS Companies

Company	Symbol	Exchange
BGS Systems	BGSS	OTC

Comments: BGS is best known for its capacity management programs, which allow a user to analyze and predict the performance of medium and large IBM computer systems and communications networks. The company's software is used not only to plan future systems but to better organize existing installations.

Company	Symbol	Exchange
Boole & Babbage	BOOL	OTC

Comments: B&B is a top SPS supplier. The company has a good growth record overall, but earnings stumbled in the first half of fiscal 1985, primarily because the firm temporarily lacked an important product for a new IBM system. However, Boole & Babbage could be back on track by late 1985. If so, it should be an excellent SPS play.

Company	Symbol	Exchange
Duquesne Systems	DUQN	OTC

Comments: Duquesne is a leading SPS supplier with products primarily for IBM and IBM-compatible mainframe computers. Customers include over 200 of the Fortune 500 industrial companies and Fortune 500 service firms, plus many universities and government agencies. The company has been extremely successful and has every possibility of maintaining its momentum. Excellent long-term growth prospects.

CHAPTER 20

Automated Data Entry Breaks a Bottleneck

If you learned of a plan to power our latest fighter planes with coal-burning steam engines, you would probably wonder how anyone could match technologies so poorly. Yet in the field of data processing we rarely give a second thought to using keyboards to laboriously enter words and numbers into computers that can often handle millions of them a second.

Actually, unlike the general public, most data processing managers have indeed given second thoughts to this front-end bottleneck and to how it can be cured. In most computer applications where information and images already exist on paper, data entry accounts for 90 percent of the costs of the project. As would be expected, an enormous market exists for automated data entry (ADE) equipment but until recently it has remained completely untapped for a lack of cost-effective products. That situation is no longer true.

BREAKING THE FRONT-END BOTTLENECK WITH OCR

The easiest and fastest way to enter text and numerical data directly into computers is by means of optical character recognition (OCR) equipment. This technology was first developed in the early 1950s but proved to be unreliable except where standardized forms could be used. In addition, it was very expensive. Although the banking industry almost immediately accepted OCR, primarily for reading code numbers on checks, other expected markets did not develop and the OCR industry sank into obscurity.

However, instead of fading away, the leading OCR companies improved their products enormously. As a result, OCR is now being accepted rapidly in markets that it failed to capture years ago and is finding many new applications as well. Problems such as high costs, frequent breakdowns, and a large incidence of errors have been eliminated. Consequently, the OCR industry appears to be in the early stages of a classic technology rebound.

OCR machines are now available for under $7,000 that will recognize standard business typefaces and misread only one character in every 10,000, well below the level of human error. Reading speeds vary with price, but most machines will work at least 10 times faster than good typists. Modern OCR machines often resemble photocopiers and contain stack loaders for automatic operation.

The key to the OCR revival is better software, a familiar story in computer-based industries. Many new systems have excellent pattern recognition programs that will make a best-guess choice in situations where a blurred or incomplete character can't be matched with certainty. Characters whose identities are in question are usually highlighted on the monitor, making them easy to spot and change if necessary.

Artificial intelligence capabilities are also available on many top-end machines. In addition to making a best guess based on pattern recognition, AI-based systems will also do a spelling check on a word containing an unknown character to see if the best guess seems correct. If not, the second-best choice is inserted, and so on. Smart OCR systems can reduce the error rate to one missed character in 300,000 and can process several thousand documents per hour.

Most new OCR systems can easily be instructed to recognize a new typeface or new set of symbols. The operator simply inserts a sample document and lets the machine ask about the meaning of any character it doesn't know. Usually, after an hour or so, the machine can handle almost anything. To the delight of the CIA, the new OCR systems can readily read

Russian, Chinese, or any other printed text whose meanings have been loaded into memory. OCR has greatly streamlined computerized translation.

A GROWING NUMBER OF APPLICATIONS

OCR now appeals to a huge market. In addition to replacing keyboard entry of routine documents in thousands of offices, the technology is becoming indispensable in many specialized industries that handle great quantities of paper. As may be expected, the U.S. Postal Service has a large commitment to OCR technology and is expected to spend millions of dollars for high-speed, multiline readers in the next few years. Department stores, utilities, and other businesses that must process mountains of payment receipts each month are becoming big OCR customers. The penny-pinching IRS has been quick to spot the potential advantages of OCR, with the result that several systems for processing tax forms are being carefully evaluated.

In addition to top-end tax-processing systems, which can recognize hand-printed characters, more modestly priced devices are also becoming available, although they require some cooperation on the part of the writers. In situations where people will use block letters and numbers to fill out orders, invoices, and similar documents, OCR is becoming very cost effective. The demand for OCR equipment that can read handprinted characters is very large and is just now being addressed.

Newer OCR devices are also being found in situations formerly reserved for bar code equipment. Warehouse order tracking can now be done with OCR machines that use TV cameras for scanning names and numbers on boxes. OCR scanning wands are often used instead of their bar code equivalents for reading invoices and similar forms.

IMAGE DIGITIZERS PICK UP WHERE OCR LEAVES OFF

In addition to automatic text entry, a major goal in the computer industry has been to enter graphical information into machines. We've reached the point where excellent graphs can be generated by most computers, but only after the underlying numbers have been laboriously typed into the equipment.

Apple Computer took graphics technology an additional step with its Macintosh, which allows users to draw pictures, graphs, and charts directly into memory with the hand-held "mouse" controller. Although the process is time consuming and is only as good as the skill of the user, the additional graphics capabilities have been identified as one of the machine's most desirable features. Many thousands of Macintosh computers have been purchased by business and technical newsletter publishers, advertising agencies, and many others for whom improved graphics capabilities are essential. The marketing information that has been collected about the Macintosh confirms other studies that indicate a multimillion dollar market for a full-function text and graphics computer system.

TO CAPTURE AN IMAGE

Placing an image directly into a computer is not as easy as hooking a TV camera to the system, although the equipment resembles video devices. Because a computer can only process digital information—combinations of zeros and ones—images are scanned as millions of tightly spaced but individual points arranged in rows and columns, each with its own locatable address. The scanning camera digitizes the image by giving each point a number according to how dark it appears. Then the mass of numbers is stored in the computer. When the image is reconstructed, it resembles a newspaper or magazine photograph with its tiny dots of ink.

Once an image has been digitized, the computer can handle it as it can any other data. Images can easily be enlarged or reduced and even enhanced for clarity by the computer. With a few keystrokes, images can be merged with text to produce camera ready publications. In fact, professional publishers and the growing number of businesses with in-house publications are placing large orders for this new technology. Image digitizers are also rapidly gaining popularity in the CAD/CAM industry where text must frequently be merged with graphics produced both on and off the computer. More general uses include the elimination of map, blueprint, and other graphics files; merging pictures with employee records for identification purposes; and using computers for facsimile communications.

The proliferation of low-cost laser printers will give a big boost to the developing image digitizer industry. Although existing dot matrix printers reproduce graphics quite well, they are no match for laser printers that can produce images good enough to publish and can do so quite rapidly. The new laser printers provide the essential output technology for image digitizers and will make the new systems much more useful.

THE BIGGEST PRIZE OF ALL

When an image containing text is digitized, the computer does not automatically recognize the characters as words, as it would if they were entered through the keyboard. Software programs that can scan stored images, recognize individual words, and convert them from graphic code into word processing code are now becoming available. Such programs will almost certainly be best sellers because they make it possible to enter huge quantities of printed text at high speed simply by scanning it. Unlike OCR devices, image digitizers can also capture whatever graphics may accompany the text.

INDUSTRY OUTLOOK

The goal of both OCR and image digitizer technology is to replace most paper files with fully controllable database records that faithfully represent the originals and may be edited with ease. That goal was set over 30 years ago and has not been forgotten by a market that has become huge. Yesterday's promise is now becoming a reality and should result in growing sales and escalating profits over the next few years.

Automated Data Entry Companies

Company	Symbol	Exchange
Datacopy	DCPY	OTC

Comments: Datacopy makes image digitizers, including models for the rapidly developing low-end market. The company also has announced a text recognition software program for scanned images that, if successful, could have a very significant impact on revenues. This is very definitely a company to review before taking an ADE position.

Company	Symbol	Exchange
Eikonix	KONX	OTC

Comments: Eikonix is a highly diversified electronic-imaging company with digitizing products for many medium- to high-end applications, including publishing, CAD/CAM, medical imaging, graphic arts, mapping, and artificial vision. The company has considerable earnings momentum and well respected products. Outlook appears excellent. A top buyout candidate.

Recognition Equipment	REC	NYSE

Comments: REI is the leading OCR supplier. The company sells systems primarily for check processing, credit card billings, currency sorting, and other large-volume jobs common to financial institutions. In addition, the company appears to be a front runner for increased IRS and USPS orders. The company has turned around dramatically after chalking up losses a few years ago and appears headed for profits.

Scan-Optics	SOCR	OTC

Comments: Scan-Optics is a leading supplier of OCR equipment for reading hand-lettered characters. The company sells and leases its products to commercial and government organizations, which use them primarily for reading payroll time cards, automobile registrations, credit card sales drafts, and similar forms. In addition, the IRS is evaluating the company's systems for reading tax returns.

Section 5

Environmental Technology: A Major Industry is Born

CHAPTER 21

We Clean Up in More Ways Than One

We have a new and sizable industry in our midst that is quietly developing into one of the most promising growth areas of our decade. The environmental technology industry came into existence almost unnoticed to service the mounting needs of manufacturers and municipalities to clean up polluted air and water and dispose of hazardous wastes. Because the press directs attention almost exclusively to the problems and controversies surrounding environmental issues, few people have noticed that solutions to our environmental problems are now being widely and profitably applied.

Substantial and long-lasting forces are propelling the environmental tech industry toward unparalleled growth. Although it takes courage to take investment positions in a new and untested area, I feel there are compelling reasons to do so.

THE OTHER SIDE OF THE NEWS

The environmental technology industry has two jobs of massive proportions that will cause it to grow from a present level of approximately $1 billion a year to at least $9 billion by 1990. The industry, fueled by a multibillion dollar Superfund passed by Congress and administered by the U.S. Environmental Protection Agency (EPA), is charged with the responsibility of cleaning up several hundred toxic waste dumps that were established years ago and now present "a clear and present danger" to the health of millions of people. Detoxifying the worst of the hazardous waste dumps will occupy the environmental technology industry through the year 2000 at a conservatively projected cost of $30 billion.

The second major job for the environmental tech industry is to serve the growing needs of commercial and municipal customers who must comply with tough air and water pollution control laws. Although most antipollution laws have been on the books for several years, until recently it appeared that legal and political challenges would render them ineffective. As a result, many of the most strict requirements were often ignored.

The controversial antipollution laws have survived recent tests and have now emerged largely intact. A broad move toward compliance is well underway and will provide additional millions of dollars in revenues to the environmental technology industry on a continuing basis.

SUPPORT FOR CLEANING UP IS WIDESPREAD

By far the most powerful force behind the growth of the environmental technology industry is overwhelming public support for a general cleanup. The public's demand for environmental reform has proved to be far more enduring that its critics anticipated in the early 1960s. Moreover, as many politicians have painfully discovered, support for a wide range of environmental issues is broadly based and crosses normal political and social boundaries. Attempts by various elected officials to weaken, mismanage, or ignore environmental quality laws have proved to be politically disastrous, resulting in scandals, resignations, and indictments.

Both government and industry leaders now agree that the need to comply with tough air and water quality standards has become a fact of life. With that agreement, the future prosperity of the environmental technology industry—and its investors—is assured.

CHAPTER 22

A Variety of Opportunities

Most environmental technology firms have specialized in terms of the materials they deal with and the industries they serve. Understanding the nature of the various specialties and the technologies they employ helps lead to good investment decisions and is well worth the effort.

HAZARDOUS WASTE DISPOSAL

Disposing of toxic substances is one of the major jobs of environmental technology companies. In many industries, the production of such substances is unavoidable, and the only safe way to dispose of them is to treat the substances and render them harmless before they escape into the environment.

We know from the experience at Love Canal and from a dozen less publicized incidents that toxic materials can't be disposed of safely in landfills. Barrels eventually corrode, burial basins leak, and the poison enters the water supply, where it is all but impossible to remove. The only cure is prevention.

Although landfills have been eliminated as disposal sites

for hazardous wastes, there are two ways in which toxic substances may be placed in the ground. One is by deep well injection, which puts wastes well below the water table. The other is to place wastes in equally deep caverns. In the latter case the materials to be stored are sometimes combined with glass or other substances in a process known as encapsulation, which is thought to stabilize the material and prevent it from escaping. Many liquid substances may be fixated—turned into solids—to achieve the necessary stability.

Sometimes "conversion" may be used with hazardous substances, that is, combining them with other materials to form a new and harmless product, in much the same way toxic chlorine may be combined with sodium to form common table salt. Occasionally the secondary material has commercial value, in which case initial opposition to detoxifying the substance usually turns to enthusiasm.

However, most toxic materials must be destroyed. One way to do this is by incineration in specially designed furnaces that reach very high temperatures and achieve 99.99 percent combustion. Oil-based materials such as PBCs will produce useful energy in such furnaces, which may partially offset the cost of the installations. Although no one wants a neighborhood incineration plant, the public definitely prefers the destruction of toxic materials to their burial, however deep.

Though many hazardous substances are detoxified by the companies that produce them, a major portion of the work is now being done off premises by specialized waste removal and treatment firms. Because dealing with hazardous materials is becoming a service industry, continuing revenues rather than one-time equipment sales are the rule.

The top hazardous waste treatment companies provide a badly needed service to a large market, and, consequently, have a prosperous future. They should appreciate substantially over the long term.

AIR AND WATER POLLUTION CONTROL

Antipollution equipment and services make up the second-largest segment of the environmental technology industry. Unlike hazardous waste firms, air and water treatment companies sell equipment rather than services because purification must be done on site. However, the sale of chemicals, supplies, and sometimes supervisory services provides additional revenues.

Most toxic substances in water may be captured chemically and removed for treatment. Quite often, toxic materials are valuable and may be reused by the company, as is true with mercury in the paper industry. Sometimes hazardous materials will settle out of the water if the water is held in quiet ponds for a short time. In other cases, the waste must be precipitated out by changing them into materials that can't remain dissolved in the water.

Biological agents are occasionally used to purify water. Bacteria may be used directly to detoxify or concentrate substances, or enzymes may be added. Although it has yet to make many headlines, the biotechnology industry is using genetic engineering techniques in serious, well-funded attempts to alter bacteria specifically for environmental applications. Quite a bit of progress has been made in selected areas, especially with bacteria that will digest spilled oil and with bacteria used in sewage treatment plants.

Cleaning up the air before it leaves a plant through the chimney is frequently very difficult and expensive. Although large particles can be filtered away with ease and small ones may be removed with electrostatic devices, toxic gases are another matter. Sometimes recycling them back through the furnace or greatly raising temperatures will lower emissions to within legal limits. However, many gases resist easy removal. Breakthroughs in this area are expected and will prove to be very profitable for the firms that achieve them. I urge you to watch for news regarding developments in this area, as it may have important implications for investors.

EMERGENCY CLEANUP SERVICES

Emergency response to environmental accidents such as spills is done primarily by hazardous waste and pollution control firms as an extension of their business. However, specialists in emergency action are coming into the picture who provide containment and detoxification services quickly enough to greatly limit the damage from accidental spills. Such firms have been of great value in recent train derailments and chemical leaks. The number of firms entering this area will undoubtedly grow significantly in the next few years.

ENVIRONMENTAL INSTRUMENTS AND CONTROLS

There is a growing market for the ultrasensitive instruments and control devices needed for air and water purification systems. Unlike similar instruments that are used in laboratories, environmental instruments must measure substances continuously and accurately, often under harsh conditions. Although many products for environmental needs are now produced by established instrument companies, several specialized firms have come into existence recently and are growing very rapidly.

An area of particular potential is the mobile environmental laboratory. Many industries that make substances by the batch don't need sophisticated testing on a continuous basis and can't cost justify their own lab. Quite a few environmental service firms now have mobile facilities to service these customers and are finding them quite profitable.

In many industries that produce toxic substances, the health of workers must be monitored periodically. As a result, several environmental instrument companies also offer medical devices and complete labs as part of their product line. Sometimes the environmental consulting industry does the monitoring for the customer.

ENVIRONMENTAL CONSULTING FIRMS

The newest and one of the fastest growing segments of the environmental tech industry is composed of professional consulting firms. Because of the complexity of environmental laws many companies have found themselves unable to plan for their own compliance. As a result, as many as half the industries with environmental problems seek outside assistance.

Environmental consultants test a company's air and water emissions, its toxic wastes, or the health of its workers, then plan the most efficient means of meeting existing standards. Many consulting firms also monitor their client's pollution control systems on a regular basis to ensure continued compliance. In some cases, consultants actually operate environmental systems for their clients.

Outside consultants are usually viewed by the courts and government agencies as impartial third parties, and are often called upon to develop environmental impact statements for manufacturers who need approval to expand or change operations. Because the technicians they employ are licensed and the accuracy of their instruments certified, the analysis done by bona fide consulting firms is accepted by state and federal agencies. As a result, consulting companies are often called in to verify compliance work already done by their customers.

In addition to their rapid growth, consulting firms with good reputations are top buyout candidates, making them even more attractive for investors. Many national engineering firms that have the talent to design and build industrial facilities and pollution control devices are prevented from entering the business because they lack environmental expertise. The quickest way to get into the game is to purchase a consulting firm. I expect many buyouts in the near future.

LOOKING AHEAD

There are several trends in the environmental technology industry that are worth noting. First, the demands for a clean environment will not disappear. If anything, there will be increased pressure for more stringent hazardous waste and pollution standards now that the public has seen some of the sad predictions of the 1960s come true. Fortunately, although the problems are no longer hypothetical, neither are the solutions.

Dealing with environmental problems effectively will require the best available technology. Landfills and other traditional systems are no longer acceptable solutions. Large, specialized firms with good scientific and financial resources are in the best position to do the work and will be kept busy for many years.

Although every one of the major cleanup technologies is imperfect and open to criticism, the dangers they present are less than the dangers inherent in doing nothing. For example, having all the contents of an old, hazardous waste dump seep into the water supply is far worse than having .01 percent of the waste escape an incineration system. Even in the press, where news articles had been sharply critical of the controversial technologies, recent articles are showing a decided shift toward the acceptance of compromise and action. Movement leading to solving problems is clearly at hand.

Who will solve the problems? A handful of companies, most of which are public, will get the lion's share of the business, the criticism—and the profits. Are we too early with our investments? I doubt it. Overall, the conditions indicate that the timing is right for the environmental technology industry, which promises to be one of the top growth areas of our times.

Environmental Technology Companies

Company	Symbol	Exchange
Environmental Systems	ESCO	OTC

Comments: ESCO is one of the leading commercial incinerators of industrial toxic wastes in the United States. Incineration is becoming the best way and, in some cases the only legal way, to dispose of many hazardous substances. This company has a good chance for increased revenues.

Company	Symbol	Exchange
Environmental Technology	ETUS*	OTC

Comments: This company, which is expected to go public in 1985, specializes in waste water management and metals recovery. A new product virtually eliminates toxic metal contamination and should prove to be very successful. I suggest you keep your eye out for news of this firm's IPO and obtain a prospectus.

Company	Symbol	Exchange
International Technology	ITCP	OTC

Comments: International Technology provides an impressive range of environmental services. The company offers design and consulting engineering services in the areas of recovery, detoxification, and stabilization. In addition, the company maintains toxic analysis and monitoring capabilities and is able to plan occupational safety and health services for industries that work with hazardous materials. The company also has emergency response contracts with the EPA to provide whatever solutions may be needed. Through a California subsidiary, the company also provides hazardous waste transport services. Offices for the company's various operations are located across the United States.

---

*Proposal symbol

Company	Symbol	Exchange
Osmonics	OSMO	OTC

Comments: Osmonics is not a pureplay in environmental technology. However, the company is a leader in membrane technology, which is finding increasing applications in water purification and pollution control. The company's membranes contain minute pores that let some substances through while serving as a barrier to others. Because membranes require no power they are especially cost effective wherever they can be used. I think this company has a good future in environmental tech and other fields. Membrane technology is a growing area with considerable potential.

Company	Symbol	Exchange
Research-Cottrell	RC	NYSE

Comments: R-C supplies air pollution control equipment and services primarily to large electric power generating plants and industrial customers. Of particular interest are the company's desulfurization systems. Since sulfur emissions are suspected of being the prime cause of acid rain, they will probably be subjected to much stricter controls in the future. The company also offers engineering and consulting services regarding many environmental technologies. In addition, the company services the nonenvironmental needs of its customers.

Company	Symbol	Exchange
Rollins Environmental	REN	NYSE

Comments: Rollins is the only large public company that obtains substantially all its revenues from providing hazardous waste disposal services to industrial customers. The company uses many of the available disposal technologies, including incineration, deep well injection, biological degradation, and solidification. Now that the EPA has declared that the burial of hazardous wastes is an unsatisfactory process, Rollins is in an enviable position. In addition to capturing much new commercial business, the company should be in line for several Superfund contracts. Excellent long-term outlook.

Company	Symbol	Exchange
Scientific	SCIT	OTC

Comments: Scientific is an integrated waste management, resource recovery, and alternative fuels company with a very practical approach to business which is very efficient. The company has been especially successful recovering the flammable methane gas that is produced naturally in landfills. The gas recovery project is the first commercially successful operation of its kind and produces electricity for the Long Island Lighting Company. Scientific is a growing company with good potential.

Company	Symbol	Exchange
TRC Companies	TRCC	OTC

Comments: TRC is a small but very promising environmental consulting company. Several years ago when the environmental cleanup movement began to gain public support, TRC was primarily oriented toward government contracts. Unfortunately, such contracts were sporadic and did not produce consistent revenues. As a result, TRC now places its primary emphasis on commercial consulting and has been much more successful. TRC should do well and is a top buyout candidate.

Company	Symbol	Exchange
Waste Management	WMX	NYSE

Comments: Waste Management is the world's largest company providing collection, storage, transportation, treatment, and disposal services. The company serves commercial, residential, and municipal customers. Although Waste Management is best known for its collection services, it is moving strongly into hazardous waste disposal. In addition to its internal efforts, the company recently bought 60 percent of SCA Services, a leading hazardous waste firm, and is expected to make more acquisitions in the future.

Other public companies with substantial involvement in the environmental technology industry include: Anderson 2000 (ANDN, OTC, air and water pollution control), Biospherics (BINC, OTC, environmental testing), Browning-Ferris (BFI, NYSE, diversified waste handling and disposal), Chemfix Technologies (CFIX, OTC, chemical waste treatment), Environmental Testing and Certification (ETCC, OTC, waste analysis and consulting), Monitor Labs (MLAB, OTC, monitoring instruments), National Environmental Controls (NECT, OTC, solid waste services), and Zurn Industries (ZRN, NYSE, air and water pollution control).

Section 6

Energy Technology: The Most Overlooked Growth Area of the Decade

CHAPTER 23

The Problem Remains

The long gas lines and fuel oil shortages of the mid 1970s are gone and largely forgotten by the general public. Although the energy *crisis* is over, the energy *problem* remains. The trouble is that for many industries and individuals, energy is still one of the major costs of doing business or maintaining a household. Not surprisingly, industrial engineers and private citizens everywhere are together budgeting billions of dollars for major improvements that will greatly increase energy efficiency.

In many cases, even small gains in energy efficiency are worth pursuing because they have a big effect on overall costs. As we shall see, in actual practice huge improvements are frequently possible using modern technology. The use of up-to-date equipment that sharply rolls back energy expenditures is spreading rapidly and is starting a growth trend of increasing strength.

The energy cost-cutting movement is not likely to level off for many years. Although oil prices have fallen from peak levels in recent months and electricity rates are also expected to soften, most industry experts are certain that the long-term trend is upwards.

Not even the threat of a new recession is likely to stall the movement toward greater energy efficiency. In fact, the revenue-slashing effects of a recession make cost cutting all the more necessary for survival.

THE CALM BEFORE THE STORM

The underlying pressures for increasing energy costs are apparent to all who take the trouble to examine our present situation. The deposits of fossil fuels that were easy to find, have been found. Those that were not only easy but also inexpensive to extract have been heavily depleted. Although geologists believe vast quantities of energy resources remain undiscovered in remote or deep parts of the earth, it is a virtual certainty that most new reserves will be very expensive to locate, recover, transport, and purchase.

The demand for energy will also begin to increase again, pushing prices to new highs. Although energy consumption dropped sharply after costs soared in the mid 1970s, most of the reductions resulted from massive worldwide conservation efforts that were very effective at correcting the worst energy inefficiencies.

Most of the low-cost, high-payoff conservation measures have now been taken. High-efficiency cars, trucks, and furnaces are in widespread use and are beginning to reach the limit of efficiency. Likewise, most buildings have been insulated as well as is practical. The least expensive fuel for each situation is now being used. In short, the fat has already been cut from the worldwide energy hog.

Meanwhile, the world economy continues to expand, as does its population. The energy saved by past conservation measures will soon be used up and the need for ever-greater amounts will resume its upward course. The process is already well underway and will become a major factor during the second half of the 1980s.

A HUGE MARKET FOR ENERGY TECHNOLOGY

By every measure the situation is ideal for a boom in modern energy technology. Most of the applications that are ideally suited for the latest generation of cost-cutting equipment and services have yet to be converted, which ensures this industry of a long and prosperous period of growth.

ONE AREA TO AVOID

One energy subsector—alternative energy—should be avoided by investors until its future is more certain. Sales of solar power, wind power, and other renewable energy equipment are very vulnerable to the loss of existing tax credits because, for most applications, they are not cost effective without them. At this writing, President Reagan has indicated that he will not support a continuation of the renewable energy tax credits that are due to expire at the end of 1985, and Congress appears unlikely to press the matter. Consequently, until the next round of energy price hikes and shortages occurs, the alternative energy industry is not likely to be profitable.

CHAPTER 24

Cogeneration Gives Two Rides for One Ticket

In almost all situations where energy is used, both heat and electricity are needed. Cogeneration equipment provides them simultaneously at enormous savings as compared to obtaining them separately.

Consider the following example. It takes about 10 gallons of oil to generate 120 kilowatt-hours of electricity. As an unavoidable by-product, 750,000 Btus of heat are produced. If the electricity is made by a utility company, the heat is lost to customers even though they paid the full cost of the fuel used by the plant to produce it. If a company needs 750,000 Btus of heat in addition to 120 kilowatt-hours of electricity, it must burn seven more gallons of oil to get it.

However, if the customer obtains energy from an on-site cogeneration plant, the company gets the heat as well as the electricity from 10 gallons of oil. The savings that occur from meeting all energy needs with 10 gallons of oil instead of 17 are so great that cogeneration plants often pay for themselves in less than two years. It is little wonder that they are gaining in popularity very rapidly.

ONCE DISCOURAGED, NOW ESSENTIAL

Today's highly efficient cogeneration equipment has up-to-date technology, including microprocessor controls. However, cogeneration itself isn't new. Cogeneration facilities were common in the United States before the National Energy Act of 1935 was passed, creating a need for public utilities by discouraging private power plants. The few cogeneration systems that survived the punitive legislation fell victim to an abundance of cheap energy, which made energy efficiency temporarily obsolete.

The energy crisis of the 1970s and its aftermath caused an almost complete reversal of the post-1935 situation. The National Energy Act of 1978, with its Public Utility Regulatory Act (PURPA), created strong economic incentives for cogeneration, including substantial tax credits. The laws also required public utilities to provide backup power for cogeneration systems and to purchase excess power from cogenerators at the cost of producing it themselves. The new laws were upheld last year in a landmark Supreme Court decision. These events, in combination with a return to high energy costs, have created growing sales for cogeneration systems.

Sales of cogeneration systems are also being boosted in some regions by local electric utilities that would rather purchase additional power at reasonable rates on the open market than construct expensive new power plants themselves. Most cogeneration systems can be constructed for $800 to $1,200 per kilowatt- hour, which is about half the cost of a modern coal-fired plant and far less than the cost of a nuclear power plant. However, in areas of declining economic growth where electric companies are threatened by cogeneration systems, their reception has been hostile despite the new laws and Supreme Court rulings.

Creative financing has come to the cogeneration industry, resulting in offers that are hard to refuse. In many cases, suppliers are able to make shared- savings agreements in which payments come exclusively from money saved because of the equipment. This results in no out-of-pocket costs whatever to the customer. Shared-savings agreements are gaining in popularity and almost always generate sales.

NEW MARKETS, NEW APPLICATIONS

The cogeneration market is expanding rapidly. In 1981, cogeneration supplied approximately 3 percent of the nation's energy. In 1983, the figure reached 5 percent. This year the level will almost certainly exceed 7 percent. The upper limit at today's energy prices is about 15 percent which, according to the Argonne National Laboratory, could be reached by 1990. That's very attractive growth in anybody's book.

Large industrial installations represent a huge market, especially those that are doubly attractive because they burn waste products produced by the customer. Cogeneration plants are ideally suited for wood and paper mills, for example, and are found even in areas such as the Pacific Northwest where low-cost hydroelectric power is readily available.

Cogeneration is also very attractive among the many industries that must produce large quantities of heat for the production process. In such cases, spent heat is directed through a special cogeneration system and used to create electricity. Heavy industrial companies of every type have found cogeneration to offer significant savings, which is helping them regain their competitive position in the world. A steel executive recently mentioned that if he could save 20 percent of his energy costs, his company would rise from the economic ashes and do some real damage to his Asian and European competitors.

Although large cogeneration installations represent a multibillion dollar market, the biggest growth is occurring in small systems for hospitals, office buildings, light manufacturing facilities, schools, and similarly sized buildings that don't use enough electricity to qualify for the lowest commercial rates. The needs of smaller energy users are being met by new, prepackaged cogeneration units that are ready to run right off the truck, require little installation, and are almost fully automated. Less than 10 percent of the huge market for turnkey cogeneration units has been served, which indicates a bright future for companies in this segment of the industry.

Cogeneration Companies

Company	Symbol	Exchange
Cogenic Energy Systems	CESI	OTC

Comments: Cogenic makes small, turnkey cogeneration systems for small to medium-sized applications. The company's products are unique because they are ready to run as they come out of the shipping crate. Cogenic recently changed its marketing strategy from selling systems outright, to offering them on a shared-savings basis, which has been very successful in the energy management business. Of course, the switch lowered revenues immediately but may pay off over the long run. I consider Cogenic to be speculative, but if the company's new marketing plan is successful, it could pay out handsomely.

Energy Factors	EFAC	OTC

Comments: Energy Factors is involved in large-scale cogeneration systems. The company currently operates four cogeneration facilities in San Diego, plus two plants that provide chilled water for air conditioning. The company has a solid standing in the commercial cogeneration market, which should grow moderately, but steadily, in the future. Energy Factors appears to be a reasonably conservative long-term energy tech investment.

Thermo Electron	TMO	NYSE

Comments: Thermo Electron manufactures a wide variety of products for energy-intensive industries. The company is a pioneer in cogeneration and makes systems of every size, from large, industrial installations to marine units and other popular small-scale products. In addition to cogeneration, the company is well regarded for its energy management systems, primarily for process control applications. Outlook appears excellent.

Ultrasystems	ULTR	OTC

Comments: Ultrasystems is a construction and engineering company with substantial involvements in cogeneration, alternative fuels, and several nonrelated projects. Although the company is not a pureplay, its fortunes swing heavily with its energy tech activities. In early 1985, the company was in a slump due to pollution concerns at two new wood-burning power plants. However, Ultrasystems is well worth watching as a recovery candidate and as a long-term energy tech opportunity.

CHAPTER 25

Energy Management Equipment: The Next Step in Energy Conservation

Many energy users have found that they can achieve substantial savings by managing their existing energy sources and uses more efficiently. Although effective energy management may call for cogeneration or other substantial improvements, quite often simply using existing equipment more intelligently is enough.

To do so usually requires that sensitive monitoring instruments be placed throughout an installation. The instruments are linked to a computer that has a sophisticated energy use and conservation program in its memory. The computer keeps track of the entire system and makes adjustments by sending signals to controls on boilers, furnaces, fans, and air conditioners.

SUBSTANTIAL SAVINGS ARE COMMON

Energy management systems save money in a variety of ways. In areas where electric rates vary with the time of day, large storage tanks may be heated at night using low, off-peak rates, then drained of their heat during working hours when it is needed. In most installations, sensors compare temperatures inside and outside buildings, then fine-tune the heating or cooling systems to maximize efficiency. Even motors can be fitted with sophisticated new devices that allow the system to match power needs with power output. The list goes on and on.

Much of the cost savings from energy management systems results from integrating the individual elements into a coordinated unit. For example, heat in a particular area can often be turned down if the lights, which also give off heat, are turned up. Likewise, excess heat from production machinery can be directed to other areas instead of heating them with the furnace. Heating for unused areas of a building can be shut off. Because temperatures tend to change slowly in a well-insulated building, heating and cooling systems can often be turned off half an hour or more before closing to save additional energy.

A modern energy management system will frequently pay for itself within two years. Suppliers are so confident that their systems will work as planned that insurance policies covering the full costs of the equipment are often included with sales agreements. Shared-savings contracts are also commonly offered.

MAJOR MARKETS

Energy management systems are becoming an important part of commercial and residential buildings as well as of industrial facilities. Most systems were initially designed for large installations, but the technology has developed to the point where cost-effective energy management is now within the reach of smaller users. The low-end market is gigantic and is yet largely unpenetrated.

Another area that is growing rapidly is sales of systems for process control facilities. It is no longer enough to have process control equipment that will create satisfactory products. Products must be made using the least amount of energy possible. Process control firms that have efficient energy management systems have an important competitive edge over those that don't.

In very small installations where purchasing an on-site computer would destroy the cost effectiveness of a proposed system, a phone link to an energy management center is frequently possible. Such time-shared environmental management systems are proving to be very effective and quite popular. Not only does the company save the price of a computer, but it gains access to a far more capable management system than its installation alone could justify.

The growth of energy management services is very positive for the energy management industry. Long-term contracts provide continuing revenue streams and profits. In addition, service contracts blunt the effects of economic changes that might otherwise cause revenues to vary widely from year to year.

THE OUTLOOK FOR GROWTH IS EXCELLENT

The market for energy management systems was $1.2 billion in 1983. It is expected to grow at a minimum 20 percent annual rate through 1990. As was true of cogeneration installations, small systems are dominating sales due to the rapid growth in small office buildings and manufacturing facilities all over the United States. Well under 10 percent of the potential market has been served, which makes this industry particularly attractive.

Energy Management and Control Companies

Company	Symbol	Exchange
Honeywell	HON	NYSE

Comments: Honeywell is a broadly oriented instruments, controls, and computer company. The company is a leading supplier of energy management systems for large buildings. (This company is also mentioned under process control in Section 1, Chapter 6.)

Company	Symbol	Exchange
Johnson Controls	JCI	NYSE

Comments: Johnson is a leading supplier of energy management systems and services primarily for large nonresidential buildings. In addition, the company manufactures process instruments for several industries and is the largest U.S. manufacturer of automotive and commercial batteries. Although the company is not a pureplay, during times when automobile sales are strong, Johnson's energy management systems add frosting to its revenue cake, which is reflected in its stock.

Company	Symbol	Exchange
Mark Controls	MK	NYSE

Comments: Mark is a leading supplier of integrated building management systems, combining energy, fire, and security functions in one computer-controlled Installation. The company is also a leading supplier of specialty valves used in process control plants. The company does very well during periods of industrial and commercial office expansion.

Company	Symbol	Exchange
Time Energy Systems	TIME	OTC

Comments: Time Energy has been very successful specializing in maximizing the efficiency of electricity usage in commercial, industrial, and institutional buildings. This company is an industry leader in shared-savings financing for energy management installations, which results in no out-of-pocket expenses for the customer. The company manages its customers' energy needs via a telecommunications link to the firm's modern, computerized control center in Houston. Time Energy has a good chance to continue its rapid growth.

Section 7

High Tech and Education Come Together

CHAPTER 26

America Updates a Critical Industry

If Benjamin Franklin had invented a time machine, in addition to his many other discoveries, he would certainly have used it to see how his young country turned out. If he were to visit our age he would probably be amazed to learn how we manufacture products, raise and harvest crops, communicate, travel, or do almost anything else. If he walked into a school, however, he would feel right at home.

It is almost unbelievable that the huge and critically important task of educating our children should be done by the least efficient of all our institutions, but it's true. We wouldn't dream of manufacturing products or practicing medicine today by using the primitive technology of 200 years ago, but the instructional process that we use in our schools has not changed significantly in two centuries. Is it any wonder that educational institutions of every kind are no longer able to cope with modern needs or have consumed a disproportionate share of our economic resources? Is anyone really surprised that the public is becoming less and less willing to support our schools and is demanding change?

SOLVING THE PRIMARY PROBLEM

Many solutions to our educational ills have been proposed, such as improving teaching standards, increasing staff size, and introducing merit pay. While all these peripheral discussions are underway, a quiet movement toward solving the real problem—low productivity—is gathering momentum.

In public and private schools, and in industrial training programs, educators in ever-increasing numbers are beginning to use new technology to increase their effectiveness. Many sophisticated electronic devices are now available that relieve professionals of many routine instructional and administrative tasks and in so doing provide them with the time to use their specialized skills more productively.

Initial resistance to technological innovation is rapidly being replaced by a clamor for more and better equipment. The people who pay the bills are beginning to see the economic benefits of obtaining technology that boosts productivity.

ONE OF THE TOP MOVEMENTS OF OUR AGE

The ultimate size and probable duration of the growing educational technology movement is staggering. Public education alone spends approximately $100 billion a year and is outranked only by national defense. When you consider that the educational productivity movement is just getting started, you can gain a measure of the enormous growth that is yet to come. When the educational needs of private institutions, industries, and the military are also considered for the United States and other Western countries, it becomes obvious that the educational technology industry will offer knowledgeable investors a great many opportunities for profits for the foreseeable future.

In the following chapters about the educational technology industry I will present several in-depth examinations of its major subfields, each of which has its own unique and particularly attractive investments. I expect you will share my enthusiasm about the growing ed tech industry and its many opportunities for profits. There is little doubt that this field will develop into one of the top growth areas of our time.

CHAPTER 27

Educational Software: The Growth Begins

At the heart of all sophisticated educational technology products are computers that rely on programmed software to perform their tasks. Software programs are so essential to the proper functioning of instructional computers and computer-based devices that in a very real sense an ed tech investment is a software investment, whether it is obvious or not.

Educational software in its many forms is presently the strongest and the fastest growing ed tech investment available. The field is still in a relatively early stage of development and has not yet become widely noticed. Fortunately, most of the companies in the industry that have the greatest long- term potential are publicly held.

A MAJOR MARKET DEVELOPS

In public schools alone, the potential size of the educational software market is staggering. For every dollar spent on a computer, eventually $2 to $3 is usually spent on software. As a result, if schools purchase only enough computers to teach computer literacy, that task alone will support a large educational software industry. When one adds the large number of computers that are being acquired for general instructional and administrative purposes, one begins to see the true size of the market that is developing for educational software. To provide each youngster in the United States with only one half hour of computer time per day would cost approximately $4.5 billion for machines and create the base for a multibillion dollar software market.

It is a virtual certainty that billions of dollars for computers and software will be found for our schools despite their present financial problems. Although the public has clearly demonstrated that it will not continue to pour money into status quo education, it has a strong history of supporting real progress and reform. In fact, in many school districts, including those in depressed areas, the public is passing budget levies targeted for computers and computer education programs. In a society where almost every worker deals directly or indirectly with high tech products, people are beginning to realize that computers will bring efficiency and increased productivity to schools at the same time that they bring a superior education to students.

INSTRUCTIONAL SOFTWARE

Most school districts spend over 80 percent of their budgets on salaries. Thus, any improvement in productivity has a big effect on the school's ability to increase its effectiveness while living within its means. If intelligently implemented, computer-aided instruction (CAI) can significantly improve teacher productivity by allowing the student/teacher ratio to increase without actually increasing the teaching load.

Newly designed instructional software is as effective as it is efficient. Lessons are given individually. Each student receives the undivided attention of a system that is far more thorough and patient than its human counterpart. Instruction is adjusted automatically for such factors as learning speed, information retention, and prior knowledge. When CAI is joined with classroom management programs, each student's progress throughout an entire series of lessons can be monitored and controlled from start to finish. The result can be an intelligently designed and administered educational program that coordinates the unique capabilities of teachers with the contributions of technology.

The best educational software is most impressive in its ability to stimulate critical thinking and is often more effective than classroom discussions. Science software, for example, usually simulates the real world by presenting puzzling phenomena that students are expected to explain. Experiments designed to uncover answers to each problem are conducted on the computer. The experiments are followed by appropriate questions. Sometimes the software guides the students to a correct conclusion, while in other cases, no useful information is given if the right experiments aren't chosen or the correct answers not given.

For more advanced problems, students usually work in teams and take days or weeks solving them. Educators have found that students who have actually developed solutions to the same challenging problems that scientists have solved learn how to think like scientists. Since real experiments are often too expensive and time consuming to conduct in every classroom, science software packages provide an excellent way to present them.

Equally effective packages exist for elementary students. One program, Meteor Mathematics, hurdles blazing objects from all directions toward a city. Each meteor contains a simple arithmetic problem. If the player quickly types the solution into the computer before the meteor hits the city, a rocket roars from its pad and destroys it in a spectacular explosion complete with sound effects. Kids love the game, which remains challenging by automatically adjusting for their increasing competence. One parent whose youngster used the experimental software remarked that his child learned more arithmetic in three days of play than he did in six years of math class.

Good CAI is an enormous contribution to the educational program, but computers can do more than instruct students. They also have the ability to test them, grade the tests, and plan the student's next lesson, all in record time. As a result, students in well-designed CAI programs spend more time at the task of learning and less time waiting for the next lesson than students in a traditional classroom. Administrators and school boards across the nation are finding that CAI can mean more students taught more subjects for less money than was previously possible. It is little wonder that highly effective instructional software is being adopted by so many school districts.

ADMINISTRATIVE SOFTWARE

Quite frequently, the first computers in a school are dedicated to administrative rather than to instructional tasks. Specialized software packages have been developed that relieve teachers and principals of routine jobs and give them time to use their professional skills to greater advantage.

Software now exists that keeps records and generates government-mandated reports regarding school attendance, student health, the progress of handicapped students, and endless details concerning federally funded programs. Other software packages can perform the time-consuming tasks of scheduling hundreds of students and monitoring their grades and graduation requirements. Packages have even been developed to administer and evaluate several routine achievement and vocational tests that were formerly given individually by school counselors.

Administrative software is rapidly being adopted because it boosts productivity in ways that school personnel find most welcome.

MAJOR SUPPLIERS

Much of the instructional software presently on the market is being created by a handful of individuals who work for small, private firms. However, the newest and most sophisticated products are beginning to come from teams of specialists consisting of computer programmers, learning theory advisers, and subject matter experts. Because textbook companies have most of the personnel already in place for such teams, and already hold copyright to considerable information, they are well on their way to taking the lead in educational software development.

Textbook companies have another great advantage over small software firms. Their products are much broader in scope and are usually designed to teach entire courses, not just single lessons. Because software packages that are part of a total curriculum are far more effective than stand-alone lessons, they will be selected over such lessons every time.

Educational software must also fit into existing school programs that consist primarily of paper, pencils, and books. Traditional education is the mainstay of textbook companies who put their familiarity with standard school programs to good use by creating and marketing appropriate educational software.

Textbook companies have an existing sales force and well-established school accounts. They also have the economic resources to develop and promote new products. All in all, textbook firms have an enormous advantage over small companies in the instructional software business.

Although textbook companies have a decided edge with instructional software, many small firms have found ready markets for well-constructed administrative packages. Large, established business software companies have also found acceptance among school districts for their suitably modified payroll, inventory, and budget programs.

INVESTMENT STRATEGY AND TIMING

As we've seen, the most attractive instructional software firms are the textbook companies that are moving strongly into ed tech. Their relatively secure financial base will provide the necessary capital to launch new products and provide a safety net for their investors while the educational software market continues to develop. Although textbook companies are not pure ed tech plays, they are pure education investments whose product mix is changing to high tech as fast as the market permits.

The educational software industry is growing much more rapidly than might have been thought likely a year ago due primarily to sharp price reductions for microcomputers. Schools are now able to buy two or three machines for a price that would formerly buy only one, and, of course, increased purchases of microcomputers for education will certainly be followed by significant software purchases.

The educational software movement is accelerating rapidly, but it is still in its early stages. Patient investors should be well rewarded for taking long- term positions before this promising industry attracts widespread attention.

Educational Software Companies

Company	Symbol	Exchange
Addison-Wesley	ADSNB	OTC

Comments: Addison-Wesley publishes textbooks and software for every educational level, from elementary school to colleges. The company recently established an educational software project which is run as an independent unit. In addition to software, the company produces a line of books on computer literacy.

Harcourt Brace Jovanovich	HBJ	NYSE

Comments: This well-respected textbook publisher appears to have been slower to develop software titles than its major competitors, but HBJ is expected to make up for the late start this year. The industry is still young enough for the company to make a good showing for itself, which should be reflected in the market.

Houghton Mifflin	HTN	NYSE

Comments: Houghton Mifflin is a major textbook publisher for all grade levels and is becoming an important supplier of educational software both for instruction and administrative applications. In addition, the company provides electronic databases for on-line financial aid, occupation, and other information.

John Wiley & Sons	WILLB	OTC

Comments: John Wiley is in a doubly strong position for the present educational situation. The company not only publishes both textbooks and software but it is heavily oriented toward science and technology. Although I expect the first major push to upgrade our schools will occur in the elementary grades, where Wiley doesn't have a strong presence, the second big push will be in high schools and colleges, which should greatly favor this company.

McGraw-Hill	MHP	NYSE

Comments: McGraw-Hill is a major publisher of books, magazines, and other information products for education, industry, and business markets. The company has a multimedia approach that includes not only software but also electronic databases for on-line information retrieval. In fact, McGraw-Hill could be more properly known as an information company rather than a publisher. Excellent long-term outlook.

Company	Symbol	Exchange
Psych Systems	PSYC	OTC

Comments: Psych Systems is a pioneer in computerized psychological and vocational testing. The company made many important contributions to the field, then experienced problems that appeared to be due to the high cost of its products. However, the concept has enormous promise and a growing following. If this company solves its problems it could be a phoenix.

Scholastic	SCHL	OTC

Comments: Scholastic is a leading publisher of educational magazines and books for children to use both in the classroom and in the home. The company also produces educational software including the most popular word processing program for younger students. Scholastic should be very successful in its efforts to take a bigger share of the growing educational technology market.

United Education and Software	UESS	OTC

Comments: United Education markets software for detailed student records, school administration, and student loan administration. The company also produces instructional software for computer literacy, word processing, medical accounting, travel reservations, and other subjects. The company operates several career schools as well. Good potential.

Other public companies with substantial or growing involvements in educational software include: Control Data (CDA, NYSE, training centers, Plato software), Macmillan (MLL, NYSE, software, textbooks), Prentice-Hall (PTN, ASE, textbooks, software, vocational schools), and Wicat Systems (WCAT, OTC, software, computers).

CHAPTER 28

Training for High Tech Jobs

As we examine the opportunities for investments in the growing ed tech movement it is important that we don't overlook the private sector. So much attention has been given to the need to upgrade our public schools that it is easy not to notice that vocational and industrial training programs have also fallen far behind the requirements of our technological society. Although public education will eventually provide the biggest market for ed tech products, much of the business for sophisticated training systems and programs is presently coming from industry, which has a critical need for highly skilled workers, and the means to create them.

The need for efficient private sector education has become enormous and will certainly grow at an accelerated rate during the foreseeable future. To understand the forces that fuel this movement is to understand its nature and develop a feel for its strength and directions. Such an understanding is extremely useful in helping spot profitable investments.

THE NEED IS CRITICAL

Most high technology industries are so short of employees with sophisticated skills that their ability to manufacture products is often severely limited. In some fields a list of experts wouldn't fill a single page. Other areas are not much better off. For example, all the people who are able to design very large-scale integrated circuits could easily fit into a high school gym. The shortage is critical in every technically advanced occupation. The growing number of firms that need such people have found that they must train their employees themselves, and in the process they are creating a large market for efficient ed tech products and services.

But the greatest need for employees with modern skills isn't among the high tech industries; it is with the huge number of rather ordinary firms that are switching to advanced equipment to improve productivity. Even for computers and many computer-based systems that have been around for years, there are still severe shortages of employees who know how to use them. The problem is getting worse as the number of companies and the variety of new devices increase.

Labor contracts that contain retraining clauses add to the demand for vocational and industrial education programs. Not only have workers successfully bargained to receive training for newly installed high tech equipment but they have secured retraining agreements for outside jobs in the event they are dismissed. This trend is receiving widespread support that crosses political boundaries and may be expected to continue. In fact, retraining efforts will probably be funded with tax dollars either directly or indirectly by means of tax incentives.

On-the-job training programs and specialized vocational schools will certainly increase in direct proportion to the modernization of business and industry. Contrary to popular belief, a strong economy can contribute to the problem because it provides profits that are often used to acquire high tech equipment.

Pressure for increased vocational training is also developing because people are remaining in the work force longer than ever before. Social Security benefits are on a sliding scale that favors delayed retirement. Even when benefits are paid at full value, the amounts are so low that many people choose to work additional years. The longer people remain in the employment pool, the more training they will require to remain up-to-date. When one adds the compounding effect that modern medicine is having by increasing the number of older people in the population, one realizes the extent of the retraining needs that are rapidly developing.

Public schools have not kept pace with the needs of business and industry and will probably not do so in the future. They are simply too slow to react to changing conditions, have different goals from the private sector, and are far too inefficient to get the job done on the required scale. Not only has industry realized the futility of relying on the public schools for properly trained employees, but so have the potential employees. High school and junior college "dropouts" are preparing themselves for high tech jobs by turning to no-frills vocational schools with intensive programs that lead to well-paying positions.

THE PRIVATE SECTOR'S SOLUTION

Business and industrial training programs are run with the same concern for efficiency that exists in other business operations. Because of the high cost of removing trained personnel from their jobs to teach or attend classes, such programs utilize educational technology to a far greater extent than do public schools.

The popularity of high tech teaching systems is also found among private vocational schools whose students wish to learn specific, marketable skills by the shortest and least expensive means possible. Such schools are experiencing phenomenal growth as students train and retrain to meet new employment goals.

Computers are an important part of an increasing number of extremely effective training systems. Often students interact directly with the computer through a terminal. In such computer-aided instruction (CAI), the computer presents various lessons and often tests the student for understanding. The best CAI is linked to computer-managed instruction (CMI), which includes software that plans a student's total program, then monitors the student's progress through each lesson until the entire course of study is successfully completed.

SIMULATORS PROVIDE REALISM

Many industrial training programs are designed to teach students how to control complex devices such as automatic production equipment. In these cases, computers are often linked with simulators that duplicate the real devices under all conditions, both real and potential. Simulators offer many advantages over training with the real equipment. Nuclear plant simulators, for example, can be programmed to present emergency situations that could never be attempted with an actual reactor. Future pilots can "crash" jumbo jets and live to learn from their mistakes.

Simulation systems are also far less expensive than most of the actual devices whose characteristics they depict. Unlike traditional classroom education, simulation training is so realistic that students are usually able to move directly from training to the job without the need for a long break-in period.

Computers are increasingly being used in conjunction with videodisk technology to form training systems that are unusually cost effective. (Please see Videodisks, Chapter 29.) In other situations videotapes are used as essential parts of more conventional self-study programs. Although such programs lack the sophistication of computer-based systems, they are nevertheless of considerble value in many applications and are widely used.

The employment needs of Western nations are changing so quickly that millions of people now expect to have several careers during their working years and look on education as a lifetime project. Meeting their needs both on and off the job are a growing number of specialized education companies that are profiting handsomely by their efforts. This industry will grow significantly and will provide healthy returns to their investors.

Industrial and Vocational Training Companies

Company	Symbol	Exchange
Advanced Systems	ASY	NYSE

Comments: Advanced Systems is a top training company with well-established products and a good reputation. The firm has an excellent line of data entry and management programs and will be introducing several industrial packages this year. The company's training uses a multimedia approach and includes traditional classroom instruction where appropriate.

Company	Symbol	Exchange
DeVry	DVRY	OTC

Comments: DeVry is a spin-off from Bell & Howell. The company owns and operates the DeVry Institutes of Technology, one of the largest systems of proprietary educational institutions in the United States. The company should do well, as the technology revolution continues.

Educational Computer	EDCC	OTC

Comments: ECC is an industry leader in low-cost, computer-controlled equipment simulators used by many branches of the U.S. military and those of its allies. In addition, the company supplies training programs and materials for classroom instruction.

General Physics	GPHY	OTC

Comments: General Physics is an important supplier of training programs, simulators, and educational centers primarily for the electric power industry. The company also provides engineering services and training programs for the U.S. military. The long-range outlook for both the defense and power industries is good, which should benefit this company.

National Education	NEC	NYSE

Comments: National Ed is the leading private vocational ed firm in the United States with over 50 schools that offer a broad range of technical subjects. The company also offers self-study programs and videodisk training systems. (See Videodisks, Chapter 4, this section.) Good long-term outlook.

Other public companies with substantial involvements in industrial and vocational training include: Control Data (CDA, NYSE, schools), Electronic Associates (EA, NYSE, training and simulation), Evans & Sutherland (ESCC, OTC, flight simulators), McGraw-Hill (MHP, NYSE, textbooks, schools), Miller Tech. (MECC, OTC, schools), and Prentice-Hall (PTN, ASE, textbooks, schools).

CHAPTER 29

Videodisks Spin Out Profits

A small but growing number of people are becoming aware of the investment opportunities that are developing because of the increased use of computers in education. Very few investors have noticed a related development, interactive videodisk instruction systems, that also has significant potential. Educational videodisk systems, which were experimental and almost unknown a few years ago, have quietly matured in recent months and are drawing increasing attention from schools, industrial training centers, and the military.

Videodisks and players have been available from RCA and Pioneer for several years as consumer items but were not successful in that application. As a result, many people have completely dismissed videodisk technology as unprofitable. In doing so, they have closed their eyes to the potential of videodisks for computer memory applications and as first-rate instruction systems.

COMPUTERS AND LASERS COME TOGETHER

Of the various types of videodisks, those that are read by lasers are the most promising. They resemble iridescent phonograph records. Each disk will store 30 minutes of high-resolution video and stereophonic sound or, alternately, 54,000 still images per side. Each image has its own identification code and can be located in under three seconds by the system's built-in microprocessor. Single images and moving sequences of any length can be addressed in any order.

Videodisk systems become powerful education tools when they are joined to small, on-board microcomputers. The resulting hybrid systems are able to present lifelike video images along with appropriate instructions. The system is very effective because it adjusts to student responses to maximize learning.

A typical program in chemistry, for example, demonstrates an actual experiment, then asks the student questions about it. The questions are not intended to simply determine if the student is right or wrong—they are designed to precisely evaluate the nature of the student's understanding of the experiment. The computer accepts each answer, then selects an appropriate response. Sometimes if the student has mastered the material, the computer moves to the next lesson. At other times the same material may be presented in a different way to clarify a point. Question and answer, back and forth, the system tutors each student patiently until the lesson is learned.

Ed tech systems using interactive videodisks are proving to be extremely efficient and effective beyond everyone's expectations. The military now uses them almost exclusively to train tank and gunnery crews. As mentioned in the previous chapter, videodisks are finding ready markets in industrial training programs as well. Both groups are very satisfied with the systems and have indicated that they will add more as additional lessons are developed. Public schools, the biggest market of all, have yet to adopt the new systems but will almost certainly do so in the next few years.

LOW COSTS GENERATE SALES

What do effective videodisk instruction systems cost? Hardware prices are now as low as $3,500 for a minimal system appropriate for a classroom, and prices will drop further as production increases. The videodisk and its software cost about $50 for the chemistry lesson discussed above.

The low prices for videodisk systems and their associated software compare very favorably with those for microcomputers alone, and in many applications are much more effective. I think the future of videodisk technology is very bright and, because they are self-contained and easy to use, they may even outsell stand-alone microcomputers for education.

Interactive Videodisk Companies

Company	Symbol	Exchange
National Education	NEC	NYSE

Comments: National Ed is primarily an industrial and vocational training company. The company uses and markets a videodisk instruction system developed with Perceptronics (see below) that appears to have considerable promise if properly handled. The company is not a pureplay in videodisks but it is a good, broadly based ed tech company. (This company is also listed under vocational training, Chapter 3, this section.)

Company	Symbol	Exchange
Perceptronics	PERC	OTC

Comments: Perceptronics is a leading maker of interactive videodisk training systems primarily for the U.S. military and, as mentioned above, has a growing involvement with industrial videodisk systems through its cooperative venture with National Education. Perceptronics has much promise as a pureplay videodisk training company.

Company	Symbol	Exchange
Wicat Systems	WCAT	OTC

Comments: Wicat is the commercial arm of the Wicat Institute, an educational think tank that has developed impressive educational computers and software as well as videodisk materials. After getting off to a bad start, the company began to rebound with deals with IBM and Control Data. Definitely a company to watch.

CHAPTER 30

Testing Systems Earn an A+

We've all seen headlines such as this: "Kids Get As and Bs in Math but Can't Do Fractions!" Usually such news follows an extensive study done at great cost by a team of specialists and is as much a shock to the schools as it is to the public.

The problem is that until very recently it has not really been practical to monitor the effectiveness of teachers and programs quickly and accurately at a price schools can afford. As a result, educational institutions lack the "bottom line" that private industry constantly faces and uses to identify both progress and problems.

HELPING SCHOOLS MEASURE PROGRESS

It isn't that schools don't give tests to measure student learning or that the test results don't contain the information schools need to evaluate their educational programs. The necessary information is available but only if the tests are subjected to laborious statistical analysis. Many schools have personnel who are trained to analyze test results but they can't do it fast enough by hand to yield useful results.

The math headline mentioned earlier reveals the nature of the problem. If a youngster takes a 100-question math test and misses every one of its 10 fraction problems, he or she could still get 90 percent on the exam. That's an "A" in anybody's class. Only if the test results are tabulated item by item will the pupil's fraction deficiency show up. If a school wishes to use the tests to evaluate the entire math program, the problem is compounded. Thousands of exams may need to be analyzed and then subjected to additional statistical studies. Without this kind of analysis, schools must often rely on guesswork to measure their effectiveness and frequently they are wrong.

THE CAVALRY ARRIVES

The introduction of computers into classrooms and school districts has set the stage for enormous improvements in the evaluation of educational programs. When computers are used with newly developed analysis software and optical test readers, exams may be analyzed in minutes. Because the analysis is available immediately, any deficiencies it exposes in the educational program can be remedied before they become more serious.

Having hard data instead of suppositions to work with has many benefits. In addition to revealing program deficiencies, good research, backed up by well- done statistical analysis, is powerful ammunition at school board meetings and bond referenda, when school personnel need to justify their requests to fund new or improved programs. Education officials who apply for federal grants have found them much easier to get and carry out when they have reliable, quantitative data. School administrators and teachers from coast to coast are finding that properly used, computerized test-grading and analysis systems pay for themselves very quickly.

HELP WITH TOUGH DECISIONS

Rational people know that some means of removing ineffective teachers and rewarding good ones must be established in our schools. Most specialists who have studied the problem know that merit pay and dismissals will only be adopted if objective criteria are used to make the necessary evaluations. The analyses made available with the new testing systems are viewed by a growing number of school officials as the bias-free evidence they need to make long- overdue improvements in their staffs.

At the individual classroom level, inexpensive test evaluation systems greatly boost teacher productivity. Instructors who formerly spent hours grading tests are now able to leave much of that job to suitably equipped computers and concentrate instead on creating more effective lessons.

During the past two years, public schools have come under increasing pressure to improve their effectiveness and become more accountable for the resources they consume. However, they have lacked the tools by which they could accurately evaluate themselves and make good decisions. Therefore, it should come as no surprise that public school officials and conscientious teachers are acquiring the new testing and evaluation systems in unprecedented numbers.

Although some nervousness accompanies the adoption of the new ed tech evaluation tools, that nervousness has not retarded sales. As one school official noted recently, "Sure, the facts and figures these systems give me are scary, but not knowing them scares me even more."

FEW PLAYERS TO SHARE THE POT

Automated test grading and analysis systems appear to be well on their way to becoming one of the fastest growing ed tech products. Fortunately for investors, there are few suppliers of such systems, which keeps the profits concentrated and the investment decisions easier to make. All the companies are public and should grow steadily for several years.

Educational Testing Companies

Company	Symbol	Exchange
HEI	HEII	OTC

Comments: HEI makes optical mark-reading equipment and other data entry devices that are used to read test responses and in other applications. Several products connect to microcomputers and are used in schools for recording attendance and other data from cards. Not a pureplay, but this company is an important supplier of OMR components that are used in many school applications.

Company	Symbol	Exchange
National Computer Systems	NLCS	OTC

Comments: National Computer is the leading supplier of large systems for educational and vocational testing in schools, universities, and businesses. In addition, the company has systems and programs for enrollment tracking, budget planning, and other record-keeping and interpretation tasks common to educational institutions. The company is able to offer rapid OMR grading and computer analysis of most of the psychological, educational, and employment assessment tests in common use. Good long-term potential.

Company	Symbol	Exchange
Scan-Tron	SCNN	OTC

Comments: Scan-Tron is the leading supplier of small to medium-sized test readers for schools. The company supplies uncomplicated stand-alone products as well as more sophisticated systems that must be used in conjunction with computers. The company's products are very popular and will undoubtedly become more so in the future. Good long-term potential.

CHAPTER 31

More Science Means More Labs

During the past two decades, America's schools neglected science and math education to such an extent that millions of otherwise capable high school and college graduates were not equipped to compete in the workplace and make contributions to our technologically oriented economy. As a result, it has become quite common, even during times of economic prosperity, to find "well-educated" people in low- paying jobs that require few skills. It is a national tragedy.

The fate of students still in school is likely to be no better than that of their predecessors unless massive corrective action is taken soon. The problem has reached such proportions that many observers believe it threatens America's ability to maintain a world leadership role in technology. It may even limit our ability to find sufficient military personnel to effectively use the sophisticated weapons we rely on for our national defense.

Fortunately, the public is becoming increasingly alarmed about America's growing technological illiteracy and is demanding swift, effective action. The call for reform is widespread and crosses social and political boundaries. As a result, Congress is responding with many new programs and allocations to improve science and math education and will almost certainly provide more of the same in the foreseeable future.

At the local level, reform is also gathering momentum. Much of the increased revenue to school districts provided by the improved economy is being earmarked for math and science. It is clear that the motivation to improve technology-oriented education is extremely strong and will prove to be durable.

SCIENCE EQUIPMENT SALES WILL BOOM

The new emphasis on science education is creating a booming market for laboratory equipment. Even in schools with no specially constructed classrooms, science instruction is possible if the necessary apparatus is available. Equipment makers are responding by putting together many different science kits that are selling very well, especially to elementary schools where the need is greatest.

Sales of expensive laboratory equipment are also turning up. Because laboratories for courses in physics and chemistry are very costly to equip, they were the most neglected when funds were short and enthusiasm for science low. Now that the pendulum has started swinging back, millions of dollars must be spent just to maintain existing programs. Millions more will eventually be allocated for the additional courses both students and parents are demanding.

ORDERS INCREASE FOR LAB FIXTURES

Although elementary science may be taught without access to a proper laboratory, most high school and college level courses must be taught in properly constructed rooms. Orders are pouring in for fixtures such as fume hoods, chemical-proof plumbing, specialty cabinets, and workstations. Even at the elementary level sales are increasing for small-scale lab installations that will fit into a corner or along an unused wall.

In many states, new regulations are being adopted regarding minimum laboratory standards, not only for all new construction but for existing schools as well. The new standards are acting in parallel with the mood of the nation, resulting in a very bright future for lab fixture manufacturers.

INVESTMENT STRATEGY

There are only a handful of science equipment and fixture suppliers, which makes it possible to participate in this growth area with only one or two investments. Appreciation should be above average over the long-term.

Science Lab Equipment Companies

Company	Symbol	Exchange
Kewaunee Scientific	KEQU	OTC

Comments: Kewaunee is the leading manufacturer of scientific laboratory furniture and fixtures in the nation. The company makes a variety of research enclosures, cabinets, glassware washers, and student workstations. Products are made in four locations in the United States and are marketed directly to schools and to industrial R&D labs. This is a solid company, in business for 79 years. I expect good, steady growth from Kewaunee.

Company	Symbol	Exchange
Sargent-Welch	SWS	NYSE

Comments: This company is a major manufacturer and distributor of an extremely broad array of scientific instruments and lab supplies, apparatus, and chemicals. The company is a leader in supplying materials for teaching secondary school science and also serves industrial, university, and government laboratory markets. Sargent-Welch will participate directly in the revitalization of our science and technical education programs. It is a good buyout candidate as well.

Section 8

Research and Development Takes On New Importance

CHAPTER 32

Profitable Changes in R&D Spending

Throughout the industrialized world a very important trend is developing in almost every business: expenditures for research and development are increasing sharply. The increases are especially significant since they are well above the levels one would expect solely from the effects of the improved economy.

When I examined corporate records from previous times of economic progress, I found R&D spending averaged 8 percent to 10 percent of revenues, whereas levels of 11 percent to 16 percent are now the norm. Federal statistics confirmed that R&D has not only recovered from its 1977 low, but that it is at an all-time high. Of greater importance is the fact that R&D spending actually rose moderately throughout the last recession, which has not happened in the past and which I did not expect to see. It is now apparent that a fundamental change is occurring with industry spending patterns for R&D. The changes present many opportunities for sizable, long-term profits for early-bird investors.

A LESSON LEARNED

It is common knowledge that during the past decade many U.S. industries lost substantial market share to competition from Europe and Japan. During the initial years the cry went out for trade barriers, subsidies, and other artificial measures to protect domestic firms. To everyone's long-term benefit, few of the Band-Aid measures were enacted, which forced corporate planners to shift their efforts toward the only lasting solution: beating the competition at its own game by producing better products and better means to manufacture those products. Although many elements are involved in the revitalization process, a central component is first-rate research and development.

I know from talking to many industry officials that R&D is now recognized across the board as the key to survival, not simply as a frill to be funded when money is plentiful. Many U.S. companies that broke with the past and invested heavily in R&D even during the recession are now doing extremely well in the world marketplace, with up-to-date products manufactured in highly efficient plants. The essential role of R&D spending in regaining a competitive edge is becoming well known. A flood of R&D activity will occur over the next several years as thousands of firms adopt what has proved to be a reliable plan for success.

INDUSTRY TRENDS

Several officials I spoke to, from companies that had cut back on R&D during the last recession, mentioned that the lack of top products and efficient manufacturing techniques had been more devastating to their firms than the recession itself. Although the recession provided a convenient excuse for their lack of foresight, these executives quietly revised their views about the role of R&D in today's highly competitive world. As a result, R&D spending will almost certainly become less cyclical in the future.

Foreign sales of U.S. R&D equipment are accelerating and will remain high for years. The U.S. Commerce Department recently lifted a long-standing ban on the export of hundreds of microprocessor-based instruments, resulting in a deluge of orders for many devices available only from U.S. firms. The economic recovery overseas and increased competition from American companies greatly adds to the demand for R&D instruments of every type.

Large orders for R&D instruments are coming from emerging industries such as biotechnology, technical ceramics, and composite materials, which didn't exist a few years ago. The new industries are at such early stages of development that they are heavily dependent on R&D equipment. This spending trend is very strong and will grow much stronger in the future.

Environmental concerns are creating another huge market for R&D instruments, as many industries establish state-of-the-art laboratories in order to ensure that they remain in compliance with tough antipollution and toxic substance laws. Analytical instruments capable of detecting materials in concentrations as low as one part per million are needed by thousands of companies in dozens of industries. The environmental market will expand rapidly at least through 1988.

Computers are having a large effect on the R&D industry just as they have had almost everywhere else. Computers make excellent collectors and processors of real-world information once they have been equipped with suitable attachments or linked directly to R&D instruments. Because of their many advantages, computers and computerized instruments have made obsolete millions of dollars of existing analytical and electronic test and measurement equipment, much of which must be replaced over the next few years.

Personal computers are also having a sizable impact on the R&D industry. A thriving and rapidly growing business exists that supplies connecting (interface) boards that allow a microcomputer to monitor many experiments and tests. Sometimes, standard business programs such as electronic spreadsheets can be used to tabulate and graph data collected by the PCs. In other cases, specialized programs are purchased from scientific software suppliers, whose field is also growing rapidly.

Renewed defense spending is providing another powerful boost to many companies in the R&D industry. Virtually everything the military purchases is done via contracts that contain the minimum acceptable specifications for each product. Contractors who do not properly establish and maintain these standards run a great risk of detection, which is often followed by stiff fines and a suspension of the right to bid on future contracts.

To ensure compliance with military specifications, defense contractors purchase great quantities of research and development equipment that they use in their labs to create products that will pass every required test. R&D companies that make instruments that can also be used in quality control applications benefit doubly from tough military standards.

The R&D industry also benefits substantially from the rapid accumulation of high tech production equipment that is occurring in most industries. Such equipment must be accompanied by related R&D instruments, which are needed for the new technologies and materials involved. As a result, all five of the major R&D subsectors presented in this section are firmly linked to factory automation, process control, and other capital-spending movements that are firmly underway.

CHAPTER 33

The Five Top R&D Investment Areas

R&D takes many forms, depending on the industry in which it is found. However, most of the suppliers fall into five major groups, each representing a separate market. Although most suppliers specialize in one area or another, a few cross boundaries. The latter companies often represent the most attractive investments, since their exposure is very broad.

ANALYTICAL INSTRUMENTS

Each substance has at least one property that sets it apart from all others. Analytical instruments are designed to exploit those unique properties in order to identify a substance's presence and its quantity. For example, atoms and molecules that have slightly different weights can be placed in a precision instrument that propels them against a slanted surface. Heavier particles strike different places on the surface than lighter ones. Some materials can be detected and measured by tagging them with dyes or other chemicals. Whatever the property, an analytical instrument exists to detect it. All of these are expensive yet essential for many industries.

Analytical instruments are also used in quality control applications where single samples that represent a large batch can be tested. The widespread need for such testing in pharmaceutical, paint, food, chemical, and similar industries represents a large, additional market for analytical instruments.

The rapidly growing environmental consulting industry also makes heavy use of analytical instruments and represents an important new market. Other independent analysis firms are springing up to meet the needs of small firms that cannot cost justify a properly equipped lab of their own.

Because analytical instruments often require the use of specially prepared chemicals and need regular servicing, they usually generate a continuing revenue stream for their suppliers. This camera-and-film relationship makes the analytical instrument industry especially attractive.

Analytical instruments are now being computerized, further increasing the investment potential of this industry by making a considerable portion of existing instruments obsolete and in need of replacement within a few years. In addition, the sensitivity of many new instruments greatly exceeds earlier models. Lastly, many new materials have been created recently that only the newest generation of analytical instruments can measure. It all adds up to a promising future for this industry.

Analytical Instrument Suppliers

Company	Symbol	Exchange
Baird	BATM	OTC

Comments: Baird supplies a variety of analytical instruments that are used in R&D labs found mainly in refineries, metal industries, academic institutions, and environmental applications. In addition, the company has used its expertise in instrumentation to create medical diagnostic devices, night vision scopes, and other products. The company has substantial foreign as well as U.S. business.

Company	Symbol	Exchange
Dionex	DNEX	OTC

Comments: Dionex is the world leader in the field of ion chromatography, which is an extremely accurate yet cost-effective technology for analyzing many critical compounds in a wide variety of materials. The company has been quite successful and even managed steady growth in income during the last recession. The market for ion chromatography is continuing to expand, which gives Dionex an excellent chance for additional growth.

Company	Symbol	Exchange
Finnigan	FNNG	OTC

Comments: Finnigan is a leading supplier of mass spectrometers, which are used to identify and measure critical levels of substances in samples. The company's products are used not only in R&D labs but also in forensic science, environmental science, and medicine. The company is a pureplay in a growing market with little competition.

Nicolet Instrument	NIC	NYSE

Comments: Nicolet is a diversified R&D equipment supplier with substantial interest in both analytical and electronic test and measurement instruments. In addition, Nicolet produces biomedical instrumentation and instrument-related computer graphic products. The company represents the broadest single investment in the R&D industry.

Perkin-Elmer	PKN	NYSE

Comments: Although Perkin-Elmer has received the most publicity recently for its excellent line of semiconductor production equipment (see also Section 2, Chapter 2), the company's roots lie firmly in the analytical instrument business, which contributes substantially to income. Of particular importance is the company's considerable lead as a supplier of computer-automated instrumentation, the fastest part of the R&D laboratory market. Perkin-Elmer should prove to be an excellent long-term investment.

Other companies with significant analytical instrument business include: Bio-Rad (BIO.B, ASE, analytical and diagnostic instruments), Kevex (KEVX, OTC, X-ray analytical instruments), Milton Roy (MRC, NYSE, mixed instruments), and Orion Research (ORIR, OTC, electrochemical instruments).

GENERAL PURPOSE, ELECTRONIC TEST AND MEASUREMENT INSTRUMENTS

Electronic T&M instruments are often confused with the highly specialized automatic test equipment (ATE) used to evaluate semiconductors and printed circuit boards on an assembly line. However, general purpose test and measurement instruments are designed to analyze fundamental electrical properties, usually in a research lab, and represent a separate investment area.

Many T&M manufacturers cross industry classifications to increase their profits by also making ATE equipment, since doing so is a logical extension of their capabilities. Many T&M firms have also been edging their way into computer-aided engineering (CAE), which is booming. The new directions T&M companies are taking will lead to increased profits in the years ahead.

Although T&M instruments are rarely used on the production line for high-volume quality control, they are used extensively for diagnosis and repair of individual electronic devices. The latter market is an important addition to the substantial sales of instruments for R&D applications.

The explosion of electronic firms and their products, coupled with high product turnover, is creating booming markets for T&M instruments, and there doesn't appear to be any end in sight.

Electronic Test and Measurement Companies

Company	Symbol	Exchange
John Fluke Manufacturing	FKM	ASE

Comments: John Fluke is the third largest supplier of electronic R&D test and measurement instruments after Tektronix (No. 2) and Hewlett-Packard (No. 1). Unlike H-P the company is a pureplay in the industry and has a broader line of products than Tektronix. John Fluke should do quite well for the next several years.

Company	Symbol	Exchange
Hewlett-Packard	HWP	NYSE

Comments: H-P is not a pureplay in electronic T&M, but it is an important factor in the industry. The company also supplies computers, electronic components, electronic calculators, and medical instruments. Future performance will be tied to a measurable extent to the success of its new RISC computers. (See related item under computers, Section 4, Chapter 2.)

Company	Symbol	Exchange
Nicolet Instrument	NIC	NYSE

Comments: A top supplier of electronic T&M instruments as well as analytical instruments. (See writeup earlier in this chapter.) Nicolet is especially attractive because of its broad involvement in R&D. Definitely a company to watch.

Company	Symbol	Exchange
Tektronix	TEK	NYSE

Comments: Tektronix is a major supplier of electronic T&M instruments, primarily oscilloscopes. In addition, the company makes several other electronic instruments and has a growing involvement in electronic design systems. Tektronix should be a prime beneficiary of the growing emphasis on R&D.

Company	Symbol	Exchange
Wavetek	WVTK	OTC

Comments: Wavetek makes general purpose T&M instruments. In the spring of 1985 the company experienced a slowdown in orders, resulting in losses. Adjustments in management and operations are underway, which may begin to have an effect in late 1985. There appears to be nothing wrong with the company's products and the market is certainly there, so I rate the company as a promising turnaround candidate.

MATERIAL–TESTING INSTRUMENTS AND SERVICES

Analytical instruments and material-testing instruments are often thought to be identical, which usually causes investors to overlook this attractive industry. Although analytical instruments will reveal the chemical composition of a material, such as the amount of graphite in new aircraft wing composites, they will not reveal the material's physical properties, such as its strength, hardness, and resistance to bending. As a result, in many industries material-testing instruments are more important than analytical instruments and receive the bulk of R&D appropriations.

Many firms cannot justify the considerable expense of a properly equipped material-testing laboratory. Instead they send samples of proposed new materials to independent testing services for evaluation. In cases where the materials cannot be moved, testing companies often send in a mobile unit or set up a temporary lab on the site. The new testing service companies represent an additional market for instruments and present an investment opportunity by themselves.

As material-testing services and instrument suppliers be

come better known, they will undoubtedly attract considerable attention. I suggest you keep your eye on this largely overlooked field.

Material-Testing Companies

Company	Symbol	Exchange
Instron	ISN	ASE

Comments: Instron was formed in 1946 after its founders, then at MIT, found the material-testing machines they needed for their research simply weren't available. Since then, Instron has become a leading source of a great variety of testing instruments. The company has sales and service centers throughout Europe, Japan, South America, and Australia, as well as the United States. The company's long-term outlook is good.

Company	Symbol	Exchange
MTS Systems	MTSC	OTC

Comments: MTS primarily supplies stress simulators for testing many materials in common use. Testing devices include stress generators for transportation equipment, seismic tables and wave generators for earthquake testing, and a variety of factory automation devices. The growth of composite materials in many new applications favors MTS.

Company	Symbol	Exchange
National Technical Systems	NTSC	OTC

Comments: National Technical Systems is an engineering and testing services company. The company designs and evaluates products for others in the aerospace, communications, defense, and energy industries. In addition, the company has a controlling interest in United Education and Software, which is listed in Section 7, Chapter 2.

Company	Symbol	Exchange
Newport	NEWP	OTC

Comments: Newport manufactures laser measurement instruments and related components, including the specialized vibration isolation tables such devices require. The company is best known for its laser systems. These are used in R&D labs primarily for material testing via holographic photography, which reveals many stresses and flaws. Holography for material testing has an excellent future and will undoubtedly replace many existing techniques in critical applications.

Company	Symbol	Exchange
Tenney Engineering	TNY	ASE

Comments: Tenney makes environmental chambers that are used to simulate the conditions a product would encounter in actual use. The company's products are becoming necessary in many industries. Tenney should grow with the R&D movement it supplies.

Another company with a substantial involvement in the material-testing industry is Wyle Labs (WYL, NYSE, material testing and electronic distribution).

REAL–WORLD TO COMPUTER TRANSLATION DEVICES

We are so used to putting numbers on every natural phenomenon—temperature 76 degrees, wind speed 10 mph—that we forget that they are artificial and misrepresent reality. In the real world, most phenomena are continuous, with no steps or isolated points. However, if we didn't put numbers on many things, we couldn't describe or analyze very much of the world. Computers have the same problem.

A small but important industry produces electronic devices that convert continuous (analog) information into the numerical (digital) information that a computer can process. These analog-to-digital devices are often far more sophisticated than may first appear and are absolutely essential in R&D labs where computers are used to evaluate real-world phenomena. When one stops to consider that the computerization of all R&D instruments—analytical, T&M, and material testing—is the next and by far the strongest growth area in the field, the important role of the analog-to-digital industry comes into sharp focus.

Real-World to Computer Connection Suppliers

Company	Symbol	Exchange
Analog Devices	ADI	NYSE

Comments: Analog Devices is the leading supplier of semiconductors and semiconductor-based components that are used for converting analog information into a digital format. The company is sensitive to cycles in the semiconductor industry but has excellent long-term prospects.

Company	Symbol	Exchange
Analogic	ALOG	OTC

Comments: Analogic makes both data acquisition products and computer equipment, primarily array processors, to process the information. (See also Section 4, Chapter 2 about array processors.) Analogic is well positioned to prosper in the high end of the research side of R&D.

Company	Symbol	Exchange
Burr-Brown	BBRC	OTC

Comments: Burr-Brown makes a broad line of data acquisition components and subsystems and competes directly with Analog Devices. Burr-Brown is a younger company than Analog Devices and is approximately one fourth the size. Consequently, B-B may offer somewhat greater potential for growth, though the downside risks may also increase. Very good long-term potential.

Company	Symbol	Exchange
Data Translation	DATX	OTC

Comments: Data Trans makes electronic products that allow IBM and other microcomputers to be used to gather and process real-world phenomena. The company is in an ideal position to benefit from the growing popularity of microcomputers in the lab.

LABORATORY EQUIPMENT SUPPLIERS

In our search for investment opportunities in the various high tech R&D industries, it is important not to overlook the less glamorous but still attractive apparatus and fixture suppliers. Companies do not purchase sophisticated R&D instruments until they have first constructed adequate laboratories to support them. Equipping such facilities with basic fixtures such as acid-proof sinks and pipes, specialized workstations, fume-proof cabinets, emergency showers, and so on is rapidly becoming a big business.

Low-tech lab apparatus and related supplies for R&D facilities are also essential and are selling very well. Scientific glassware and glassware washers, incubators, balances, timers and regulators, power supplies, and similar gear is needed in great quantities.

As I noted in the section on educational technology, additional revenues for the laboratory apparatus and fixture industry are coming from schools and universities that have started to make long-overdue improvements in their science and math programs. The potential size of the educational market is staggering. Although the ed tech area is developing more slowly than the industrial markets, its long-range potential is well worth considering when investment decisions are being made. An investment in the lab apparatus and fixture industry could give you a double play for your money.

Laboratory Apparatus and Fixture Suppliers

Company	Symbol	Exchange
Heinicke Instruments	HEI	ASE

Comments: Heinicke produces environmentally controlled apparatus for general laboratory use. Products consist of laboratory ovens, incubators, sterilizers, liquid baths, and other common lab devices. Through a subsidiary, the company produces and repairs parts for gas turbine engines. Although not a pureplay, Heinicke is unusually well positioned to profit from the growth in R&D labs. I would expect good, solid, long-term growth from this company.

Company	Symbol	Exchange
Kewaunee Scientific	KEQU	OTC

Comments: Kewaunee Scientific was discussed under educational technology, Section 7, Chapter 6, about lab fixtures. Kewaunee is well positioned to profit from the growth in R&D as well.

Company	Symbol	Exchange
Sargent-Welch SCI	SWS	NYSE

Comments: Sargent-Welch was also discussed in the ed tech section. Like Kewaunee, this company's market extends into commercial R&D, which should provide additional revenues.

PLEASANT SURPRISES?

The R&D instrument and apparatus industry looks extremely attractive. It couldn't be more ideally suited for meeting the tough competitive needs of industries worldwide. I expect the R&D sector to appreciate substantially for the next several years. Moderate growth should even occur during economic slowdowns.

Speaking of economics, we have a good chance to have several more years of unprecedented growth, albeit with the temporary setbacks that are common to all trends. The American industrial machine is stronger and far more adaptive than its counterpart in any other country, and is just now beginning to demonstrate its new capabilities. These new capabilities, which depend heavily on top-notch R&D efforts, will be increasingly apparent over the next several years and should greatly surprise even the optimists. Fortunately for investors, prices for almost all U.S. industrial companies do not reflect the extent of the growth I feel is likely to come. Certainly the central role of R&D has yet to be generally recognized in the investment community.

Section 9

High Tech Suppliers: Huge Rewards with Reduced Risks

CHAPTER 34

The New Winners in the Technology Wars

In the early days of a new technology, when competition is low, manufacturers and stores have an easy time chalking up record profits. As the technology matures, however, competition enters the picture which often leads to vicious price wars and industrywide shakeouts.

Shakeouts are compounded by frequent introductions of new products and the rapid obsolescence of old models. Profits become skimpy and investment values erode. Even the "winners" can be damaged so severely during a shakeout that they don't represent good investments.

BUT THERE *IS* MONEY TO BE MADE

All seasoned investors know that there are profits to be made from any major event, although perhaps not directly. The lower prices and profusion of goods that hurt the manufacturers and retailers are a boon for consumers who react by purchasing high tech products in unprecedented quantities.

Intuitively, we sense that there is good money to be made as prices fall and sales accelerate, but with a shakeout in progress, we have to take a lesson from ole Brer Fox and find another way to get what we want. We must stay off the battlefield and invest instead in companies that benefit from the vastly increased numbers of products being produced.

In the area of microcomputers and other consumer electronic products, the electronic distributors and connector companies are rapidly emerging as the ultimate winners in the battle for profits. Of course, no business is a bed of roses, but unlike the end-product manufacturers that are caught directly in the shakeout, distributors and connector companies often find it possible to maintain good profit margins while prices for finished goods decline.

PROFITS FROM DEVELOPMENT STAGE INDUSTRIES

In the biotechnology industry we also have a situation where suppliers make excellent profits, in this case, years before products will be ready for marketing. The lucrative biotech supply industry is an outstanding sideline play for investors who wish to participate in the explosive growth of the biotechnology industry.

Staying away from companies in the development stage or in a shakeout is always good advice. However, if you carefully invest in their suppliers, you can make a great deal of money while keeping your risks under control. It's just the sort of situation ole Brer Fox would love.

CHAPTER 35

Profits with Safety for the Distributors

Distributors are in a very enviable position during shakeouts in high tech industries. The more manufacturers there are, and the lower prices go, the more products are created. Distributors win at both ends of the production cycle. Some distributors supply the electronic components that go into those products, and others handle the finished goods for the manufacturers. Unlike the manufacturers, who are limited to their own line of goods, distributors can handle anyone's products. If the products are popular, distributors rarely care what they are or who supplies them.

DISTRIBUTORS WON'T BE BYPASSED

Distributors are sometimes seen by investors as superfluous middlemen in the electronics industry and therefore of questionable investment value. Critics of the industry point out that both manufacturers and retailers are often able to obtain products directly from the producers. Although direct sales do occur, and may continue with some items, the practice generally seems to be on the decline.

In some cases involving consumer electronic products distributors do get bypassed, but they can still reap handsome though indirect profits from them. Often the most money to be made with such products is with accessories such as peripherals and software that are needed in huge quantities. Just as stereo hi-fi manufacturers provide a base for the far more profitable record industry, so do computers, VCRs, and similar products support a vast accessory and supply industry. The underlying technology itself is often the smallest and least desirable part of the prize.

The biggest advantage distributors have is that manufacturers of components and finished products are poorly suited for the wholesale business and usually do very badly at it. For the first year or so following the introduction of a new line of products, manufacturers may supply customers directly because distributors are often slow to take on unproved merchandise. As demand for the products increases, however, distributors become more willing to enter the picture. At that point, most producers bail out of the wholesale business with a hearty "good riddance."

In most cases, customers are quite happy to have the services of top-notch distributors. Computer stores, for example, rely heavily on distributors for technical support that includes training sales personnel to use and promote the various types of merchandise they carry. Likewise, manufacturers rely on electronic component distributors to show them how to design, assemble, and test the finished goods that contain the products they handle. All customers benefit from shared advertising plans, market research, and product evaluations provided by on-the-ball distributors who know such services are badly needed and highly valued.

DISTRIBUTORS MODERNIZE AND PROSPER

The distribution business is changing rapidly, especially that part of the industry that supplies basic components. Although the largest component customers frequently want bare parts, there is a clear trend toward value-added or partially manufactured items such as plugs with the wires already attached. In an increasing number of cases the distributors do the necessary preassembly work, bringing in additional revenues. Value-added work also helps solidify accounts because customers have more to lose if they switch suppliers.

Another area that is undergoing rapid change is semiconductor distribution. As I mentioned in Chapter 10, on custom and semicustom chips, distributors are adding CAE capabilities as fast as they can in order to serve the expanding demand for the new products. Here again we find value being added to generate more business.

AN ENVIABLE POSITION

Distributors enjoy a degree of safety in their business that is not found with most manufacturers or retailers. Because they have many suppliers, products, and customers, no failure or bankruptcy with any one of them is likely to be devastating. Distributors buy from the winners and sell to the survivors, whomever they may be. In addition, distributors usually protect themselves be securing buy-back agreements from manufacturers of unproven products, and they only stay with merchandise they can turn over at least 8 to 12 times a year.

Distributors are increasingly adopting high tech products to streamline their operations and boost profit margins. New order-tracking tags and computer programs allow distributors to see at a glance what products are selling, and where.

SMALL DISTRIBUTORS ARE ESPECIALLY ATTRACTIVE

Successful small distributors of new products are frequently attractive buyout candidates. Larger firms tend to be slow to take positions in top-selling products. When the big firms find themselves behind the eight ball, they know they must establish strong beachheads quickly or be shut out altogether. In a recent example, the industry giant, Arrow Electronics, snapped up High Technology Distribution right after acquiring Computer Products Supply and Computrend. Other electronic distributors are also likely to decide that buying an existing firm is the best way to increase their involvement in new high tech products.

Buyouts also occur when a big firm wants to serve a particular geographic area that is presently handled by a smaller company. When Silicon Valley and Route 26 started to boom, it triggered many acquisitions of smaller firms by large distributors who were caught off guard by the unexpected growth. I expect small firms with strong ties to the developing high tech industries in Texas, Minnesota, and Oregon will now be acquisition targets.

INVESTMENT OUTLOOK

Distributors of high tech products should perform well over the next few years, providing the economy remains reasonably strong. Although the industry is sensitive to the economy, lacks glamor, and isn't the focus of much attention by the financial press, it should make attractive profits for a longer term than is true for many of the companies whose products they handle.

Distributors

Company	Symbol	Exchange
Anthem Electronics	ATM	NYSE

Comments: Anthem is a distributor of semiconductors and semiconductor systems, primarily in the western United States. The company is particularly noteworthy because it uses very aggressive pricing on commodity semiconductors to get orders for products with high average selling prices. The strategy permits Anthem to keep the number of product lines low and profits high. The company has been quite successful.

Company	Symbol	Exchange
Avnet	AVT	NYSE

Comments: Avnet is the world's largest distributor of semiconductors, connectors, and electronic components. The company may be expected to do well during times of economic expansion and represents the most conservative investment in the industry.

Company	Symbol	Exchange
Handleman Company	HDL	NYSE

Comments: Handleman is the largest distributor of prerecorded music in the Untied States and Canada. The company uses a sophisticated computer ordering and distribution system that permits Handleman to monitor the performance of each product it handles in each store it serves. The company has found that microcomputer software and books are natural extensions of its business and has successfully added them to its product lines. Although the company is not a pureplay, technology products may have a significant impact on revenues in the future.

Company	Symbol	Exchange
Micro D	MCRD	OTC

Comments: Micro D is essentially a pureplay in microcomputer products. After getting off to an excellent start, the company experienced a number of difficulties. At this time, the company is making many changes, both in personnel and marketing strategy, in an attempt to get back on track. The company has good turnaround potential. A glance at an up-to-date stock chart will quickly reveal if the new programs are having the desired effect.

Company	Symbol	Exchange
Pioneer-Standard	PIOS	OTC

Comments: P-S distributes consumer electronic and computer products. In addition, the company handles electronic components and has a small involvement with instrument sales and repair. Most of the company's business is conducted east of the Mississippi. However, the company is well established in Minnesota and Texas, two fast-growing high tech markets.

Company	Symbol	Exchange
Western Micro Technology	WSTM	OTC

Comments: In 1984, Western Micro was the fastest growing publicly held distributor in the United States. The company handles finished products and a full range of semiconductors, mostly in the rapidly growing electronic centers of the West Coast. A new CAE design center is boosting sales of custom and semicustom chips. I believe the company is a top buyout prospect.

Other public companies with a significant involvement in the distribution industry include: Arrow Electronics (ARW, NYSE), Diplomat Electronics (DPLT, OTC), DKM Electronics (DKME, OTC), Ducommun (DCO, ASE, not a pureplay), Marshall Industries (MI, ASE), and Wyle Labs (WYL, NYSE, not a pureplay).

CHAPTER 36

The Connector Industry Connects with Growth

Some investors in high technology companies may be tempted to turn up their noses at connector manufacturers. There is little doubt that plugs, cables, and sockets are as dull as dust next to computers, lasers, and other devices that have more sizzle. However, for people who aren't too proud to take advantage of a nice profit opportunity anywhere it may be found, the connector industry frequently pays off in a big way.

Actually, connector companies are all the more attractive precisely because they make many investors yawn. Individual investors frequently don't pay much attention to these stocks until they make a good move, which gives early birds who have done their homework a little extra time in which to take positions. When the stocks do come to everyone's attention, nice profits may then be taken.

CONNECTORS ARE GOOD BUSINESS

The connector industry profits from growth and advances in many fields, including computers, telecommunications, medical devices, aerospace, instruments, consumer electronics, and others. Because connectors are used in such a wide variety of products, profits are somewhat protected from sharp drops during economic slowdowns. However, since the connector industry tends to move smoothly and predictably with the economy, in knowledgeable hands its cyclical nature can be one of its most profitable characteristics. But more of this later.

Not only is the number of products that use connectors going up but the number of connectors per product is also increasing. With only a few exceptions, the more sophisticated the technology becomes, the more connectors are needed. Thus, the present explosion in the number of products that offer advanced features is of double value to the connector industry. With the average number of connectors per product increasing, the connector industry may grow even faster than the industries it serves.

Although the number of connectors per product is increasing, connectors still tend to make up only 1 percent or 2 percent of a product's total cost. As a result, connector suppliers are usually able to raise prices modestly year after year, even when business is slow. They are rarely subjected to the same pressures to cut prices that are so common among suppliers of more expensive components.

PROFITS ARE BEING PUSHED UP

There are several developments occurring in the connector industry that should have a positive impact on profits. Many customers are starting to show a preference for value-added products, such as cables with the connectors already attached, where formerly they would have purchased the components separately. Value-added products carry a higher margin than do individual components.

A boost in profits is also coming from new computer and telecommunication devices that require connectors that are more sophisticated than their older counterparts. Fiber-optic links, "zero insertion force" connectors, and products that cut down on electromagnetic interference (static) are all in demand. The new connectors are also high-margin items.

Curiously, for an industry with a worldwide market that is expected to exceed $9 billion this year, there is relatively little competition. Only a handful of companies are doing 90 percent of the business. The market for connectors is so diverse that most companies have settled into separate areas and lock horns with each other far less frequently than one might expect. Foreign companies have largely been content to remain in their respective countries, leaving the huge U.S. market primarily to domestic firms.

THE INVESTMENT CONNECTION

The connector industry has much to offer the investor, especially at turning points in the economic cycle that may be occurring again soon. Because increased demand for electronic products usually precedes an economic upturn, the connector industry can be a good early bird recovery investment. Likewise, demand for products drops prior to an economic slump, with a predictable effect on connector sales.

Due to their close relationship with the health of the economy, connector stocks are a good way to play the economic cycles. Institutional investors have been using connector stocks for years to help point out cyclical turns and then to make money from the trends. It is high time individual investors make the same use of this attractive investment technique. Individuals who understand the prudent use of options can play the leading connector stocks at both ends of the cycles.

The connector industry also offers investors an excellent opportunity to participate broadly in the rapid growth of the electronics sector with somewhat greater safety than might be true of more direct involvement. The recent microcomputer wars are a good example of how connector companies can prosper while their customers beat each other to a profitless pulp in the marketplace. Investors who played the microcomputer boom via the connector industry made good money while many people who purchased microcomputer stocks lost heavily.

For the past 15 years, connector stocks have consistently outperformed Standard & Poor's 500 index. This trend may be expected to continue for at least the duration of this economic cycle. Returns on equity for the connector industry have averaged 20 percent in recent years. All in all, the connector industry holds promise for careful investors who understand it and its unique relationship with the economy and the electronic revolution.

Connector Companies

Company	Symbol	Exchange
AMP	AMP	NYSE

Comments: AMP is the giant of the connector industry. The company's product mix has shifted dramatically in recent years from an electrical to an electronic orientation, with three fourths of the company's sales now going to the computer, telecommunications, and associated industries. In addition to connectors, AMP offers special tooling and value-added products that are making an increasing contribution to earnings. AMP always does well during periods of economic growth.

Company	Symbol	Exchange
AUGAT	AUG	NYSE

Comments: Augat specializes in connectors for semiconductors and related microelectronic parts. The company's products are most commonly used inside rather than between electronic devices. Augat's sockets and other products are used extensively in computers, telecommunication devices, and similar equipment. Augat's long-term prospects appear excellent.

Company	Symbol	Exchange
MOLEX	MOLX	OTC

Comments: Molex is a leading supplier of connectors, connector cables, and associated products used with computers, computer peripherals, and many consumer electronic devices. A substantial portion of revenues come from foreign sales that increase substantially during times when the U.S. dollar declines in value. Because the company's focus is on the most rapidly growing electronic products, Molex may well outperform much of its competition.

Other public connector companies you may wish to investigate include: Burndy (BDC, NYSE, connectors for electric utilities and electronics), Matrix Science (MTRX, OTC, connectors for hostile environments), Robinson Nugent (RNIC, OTC, mixed electric and electronic products), and Thomas & Betts (TNB, NYSE, mixed electric and electronic products).

CHAPTER 37

Biotechnology Supply Companies: The Smart Way to Biotech Profits

The biotechnology industry has a few successes it can point to with pride, but for the most part, it occupies first place in the disappointment line on Wall Street. The industry received enormous publicity long before products were ready for the market in sufficient numbers to provide the companies, and their investors, with a profit. Although a handful of bright stars are emerging, the industry is still largely in the development stage and substantial returns are years away.

Besides the difficulties inherent in unraveling the complexities of genetic material, there are formidable legal obstacles to bringing any drug or medical diagnostic product to market. Even after a new discovery has made it through the lengthy development stages, it must confront the burdensome regulations of the Food and Drug Administration, which can tie it up for years in extensive testing often costing millions of dollars. When the product is finally released, it is no more assured of success than is any other product. In fact, quite a few biotech developments have run the legal and financial gauntlet only to flop in the marketplace.

The prospects for agricultural and veterinary biotech companies are a little better than are those for firms that deal in products for people. If you wish to make a direct investment in the biotech industry, I urge you to stay in the ag/vet area where the regulation process, and the costs, aren't so crippling.

Eventually, of course, many investors will make a great deal of money in the biotechnology field. The industry is maturing and will live up to its promise as the process continues. However, because of the difficulties and the delays mentioned earlier, many more investors will either lose money or will wait so long before profits are realized that they would have been much better off taking positions in other areas.

Although biotechnology isn't the most attractive investment available, there is a clever way to make money from the explosive growth of the industry. Instead of walking into the twilight zone with a front-line company, you can invest in a biotech supply firm. Such firms take few of the risks common to their customers and quite frequently get most of the profits.

AN ENVIABLE POSITION

The biotechnology supply industry is the sole source for the specialized instruments and laboratory materials that are essential for the research-dependent biotechnology firms. The supply companies have found immediate profits and relative safety in an otherwise shaky field. However, because they lack the glamor of the more newsworthy biotech companies, they have gone largely unnoticed by the financial press. As a result, several of them represent an attractive way to play the rapidly expanding biotechnology industry, an industry whose impact on our lives may someday exceed that of microelectronics.

BIOTECHNOLOGY SIMPLIFIED

Biotech supply firms and their products are best evaluated from an investment standpoint if one has a fundamental understanding of the industry they serve. A brief overview of the field follows, which should be helpful as you examine biotechnology investments of every type. The emphasis is on the major areas and concepts that are almost always poorly understood and yet are essential to understanding biotechnology and its many investment opportunities.

GENETIC MANIPULATION

All complex plants and animals consist of countless individual cells that contain all the instructions that are needed to enable them to perform their life functions. The instructions are built into long strands of DNA molecules, which resemble twisted ladders. Although these strands contain only four basic kinds of molecules, they may be arranged in millions of different patterns. Each unique pattern is a separate instruction for creating an essential cell product and is known as a gene. The long ladderlike DNA molecule may be thought of as a double strand of genes, each of which contains a specific instruction, or blueprint, for a cellular function. When the cells duplicate themselves as the organism grows, they also duplicate their DNA.

The makeup of DNA and its genes is somewhat like this sentence. Or rather I should say themakeupofDNAanditsgenesissomewhatlikethissentence. Each letter of the last line corresponds to a molecule in the DNA strand and each word, which corresponds to a gene, is really just a particular pattern of those molecules.

If we know the patterns to look for and what they mean, the job of deciphering a DNA strand is easy. But if the strand looks like this—atgcttaacccgaatcgtgaccatgcataa—and continues for a long distance, the job takes on a different character. You don't know how long each word (gene) is or where one begins and another ends. Of course, even if you think you have identified a gene, you need to determine what it means. Add the additional problem of not being able to actually see the strand or manipulate it directly because it is too small, and you begin to appreciate what cell biologists are up against.

Not surprisingly, thus far, no one has succeeded in deciphering the codes for more than a few genes, even though scientists know which molecules make up DNA strands and how to hook them together. Some successes have been made, however, with impressive results. For example, Genentech cracked the code for the insulin gene found in the DNA of human pancreas cells, then succeeded in splicing together that gene with the DNA of a particular type of bacteria. The bacteria thereupon manufactured insulin as a normal part of their activities. Similarly, human growth hormone and interferon are now being produced by living materials with altered DNA.

The Genentech achievement with insulin was a triumph in isolating, then recombining, genes from different organisms—recombinant DNA technology. However, the real goal of DNA research is to completely decipher the genetic code and achieve true genetic engineering, which is the ability to create DNA with made-to-order genes that will produce custom plants and animals.

True genetic engineering will take many years to accomplish because the DNA code is incredibly complex. Some genes only work in combinations. Others exist as fragments and are fully formed only when the DNA strand coils back upon itself, placing the fragments next to each other. In addition, the process by which many genes are switched on and off is poorly understood.

Quite obviously we have not really achieved genetic engineering, despite all the chest pounding to the contrary. We also have a great deal yet to learn about recombinant DNA technology which, while impressive, is even less sophisticated than genetic engineering. As a result, most genetic manipulation consists of "simply" fusing together whole cells that have genetic codes that we would like to have in a single organism.

Cell fusion is the technique used to create monoclonal antibodies, which are of great value in diagnosing and treating many diseases. Antibodies are compounds produced naturally by certain cells in the body. They attach themselves to foreign substances such as invading bacteria or cancer cells. Once they are attached, they help identify the harmful substances so that they can be located and destroyed by the body's defense system. Since antibodies are specific to particular substances, they can also be used to detect the presence of those substances in a lab test. For example, if a few drops of herpes antibody are dropped into a blood sample taken from a person with the disease, the sample will turn color.

Unfortunately, the cells that produce antibodies do not live outside the body, so growing cultures of them for antibody production isn't yet possible. On the other hand, certain cancer cells do grow well in laboratories but they don't form antibodies. The solution, as you've undoubtedly guessed, is to fuse the two kinds of cells and create a hybrid that can be artificially cultured. Monoclonal antibodies are produced by one (mono) type of hybrid cell and its identical offspring (clones).

BIOTECHNOLOGY SUPPLY

As we've seen, biotechnology is a diverse area, but the various fields of biotechnology all share one important characteristic: they all depend on—and heavily support—expensive, long-term research and development programs. These R&D programs require huge quantities of specialized products that are available only from biotechnology supply companies.

Perhaps the most exotic products developed by the biotech supply industry are so-called gene machines, which make it possible to automatically build short strands of DNA with whatever code is desired. Related machines accept DNA from living materials and take it apart, molecule by molecule, while recording the code. Researchers can then make duplicates and attempt to find out what the code means.

Interestingly, gene machines are often accompanied by computer programs that use cryptographic analysis techniques borrowed from the CIA and the National Security Agency. As one biotech researcher remarked, "Codes are codes and this cloak-and-dagger stuff cracks DNA secrets better than anything else we've tried."

Biotech supply companies also sell fermentation equipment that is used to grow microorganisms for recombinant DNA and monoclonal antibody research. Fermentation devices are also used in production plants to grow large quantities of the living materials that make biotech products for the market.

Specialty chemicals are also used in large quantities by the biotech industry, which must obtain them from the supply companies. Tons of very profitable chemicals are sold each year to front-line biotech firms, which require them in ever-increasing amounts.

Biotech supply companies are in a unique position. They profit from the explosive growth of the biotechnology industry, but they do not share the usual high degree of risk common to their customers. Although such customers must wait years for their own research to pay off, their suppliers tend to make immediate profits. For every customer that doesn't survive the long wait to profitability, it seems that two new companies spring up to take its place.

The biotechnology supply industry is not without risks, but it is in a far better position to make immediate profits with more safety than its customers. It is a very attractive area and will undoubtedly become even more so as the biotech field continues to expand.

Biotech Supply Companies

Company	Symbol	Exchange
Applied Biosystems	ABIO	OTC

Comments: Applied Bio is the world's leading supplier of gene machines, which are used to create genetic materials and decode natural DNA. In addition, the company supplies genetic materials and other biotech research products. The company's products have received unprecedented acceptance. At the present time, the company has little effective competition. Although the clear sailing won't last forever, Applied Bio has a big head start and will be difficult to catch.

BIO-RAD LABS	BIO.B	ASE

Comments: Bio-Rad is a leader in separation technology, the means by which biological materials and chemicals are isolated for identification. In addition, Bio-Rad manufactures diagnostic kits and analytical instruments for basic research and clinical medicine.

New Brunswick Scientific	NBSC	OTC

Comments: New Brunswick is best known for its excellent line of fermentation equipment, which is needed to grow the microorganisms used in research. Of greater importance, and often overlooked, is the critical role fermentation systems play in the large-scale production of biotech products. Because the biotech industry is maturing, the need for production systems will increase steadily.

Sigma-Aldrich	SIAL	OTC

Comments: Sigma-A is a major supplier of biochemicals and other materials that are used in great quantities by the biotech industry. The company has over 40,000 products that can be shipped within 24 hours from locations around the world. This company has been very successful and should continue to grow.

Other public companies with substantial involvements in biotech supply include: Collaborative Research (CRIC, OTC, biochemicals), Enzo Biochem (ENZO, OTC, enzymes and filters), Millipore (MILI, OTC, filtration systems), Nicolet Instrument (NIC, NYSE, analytical instruments), Pall Corp (PLL, ASE, filtration systems), Pharmacia AB (PHABY, OTC, separation technology and pharmaceuticals), and Vega Biotechnologies (VEGA, OTC, gene machines and materials).

Section 10

Special Niches Yield Big Profits

CHAPTER 38

Opportunities Are Sometimes Where You Least Expect to Find Them

Most of the investment opportunities I've presented so far have involved technology movements of major proportions that span many fields and have far-reaching impacts. Those movements deserve their featured status because over the next several years, they will return many millions of dollars to countless investors who had the foresight to learn about them and take wise positions.

In our search for profits, however, it is important not to overlook smaller and less newsworthy high tech areas that have unusual promise. In such areas, one can often find very successful companies that will appreciate handsomely. Searching for them can be enjoyable as well as financially rewarding.

SEEING WHAT OTHERS MISS

Often the best place to find top values is in areas that other investors expressly avoid. I'm not advocating a contrary-opinion approach because that strategy is often done almost as an act of faith and without much thought. However, it can be very rewarding to play the devil's advocate and attempt to construct a case in favor of an out-of-favor industry or company. Doing so will often reveal a dark cloud's silver lining.

A good example of an industry that has very profitable subfields that are almost totally ignored is radiation technology. The field has such a negative image that if you mention that it interests you, people are likely to stare incredulously. However, they will be thinking about Three Mile Island, not low-cost sterilization of foods and medical instruments. They will be thinking about the huge losses in the nuclear power debacle, not the profitable adaptation of computerized X-ray machines to industrial inspection.

Another industry that "everyone knows" is a sure loser is computer retail stores. There is no doubt that the industry is presently in a bloody shakeout resulting in huge losses and few profits. But after every battle a few winners always emerge, dust themselves off, and claim the spoils of victory. Fortunately, the front runners can be spotted in the final battles and their stocks can be obtained at bargain basement prices. Since well over three fourths of the market for microcomputers has yet to be served, the winners will prosper and are well worth finding. To hear most investors talk, though, you'd think the area wasn't worth a second glance.

Frequently, new opportunities will come to your attention if you allow your natural curiosity a free rein. Recently, a friend's son, who works in a defense plant, showed me a new security pass that opened my eyes to a developing industry I hadn't previously investigated. The pass wasn't the usual unflattering mug shot and fingerprint. Instead, it contained a microchip that activated special security devices and gave detailed information about the person to a verification system. I was amazed by the high tech pass. When I followed up on my curiosity I uncovered many attractive investments in the security industry.

A similar situation occurred during a visit to a local print shop at the end of a long day. I'd just come from a series of visits to high tech companies and had been surrounded by state-of-the-art equipment. By contrast, the print shop made me feel that I'd stepped back in time to the age of steam engines and cast iron. But although the presses resembled antique locomotives, they were only a few years old. I was struck by the fact that the industry was a sure market for automated equipment, if any existed. A week later, I knew it did exist and represented a multimillion dollar industry in the making.

SELF CONFIDENCE IS ESSENTIAL

If you look at the history of investing you will find the big profits have always gone to people who thought for themselves and stuck to their plans. When you spot an opportunity that is out of the investment mainstream or is out of favor, you will need to examine it, then follow through on your own.

Unfortunately, original thinkers can expect little support from other investors, who are often unreceptive to new ideas. They may even try to convince you to change your mind. Before you listen to them, please remember that more money has been made by investors that followed their own instincts than by any other group.

CHAPTER 39

The Bright Side of Radiation Technology

The nuclear power industry has been a fiasco, marked not only by the accident at Three Mile Island but also by scandalous losses that have cost investors billions of dollars. The threat of nuclear war and the publicity about the horrible effects of radioactive fallout have added to the overall negative image that surrounds anything or any company that deals with the phenomenon of radioactivity. The negative feelings are so strong that people assume without question that the whole industry is bathed in red ink and has no future whatever.

As most experienced investors have learned, however, challenging widely held assumptions can pay big rewards, because the process often uncovers opportunities not noticed by others. Nowhere is this more true than with the multifaceted radiation tech industry, where many existing applications are booming and others with outstanding potential are just coming into existence. This industry is doing far too well and has too much potential to escape attention for long.

UNDERSTANDING WHAT OFTEN CAN'T BE SEEN

Radiation, as the term is commonly used, consists of tiny particles or bundles of energy given off by atoms. The particles or energy packets are of different sizes and they behave differently as well. Since most types of radiation can be generated artificially by one means or another, radiation can be put to use. The type of radiation selected depends on the job that needs to be done.

For many tasks, such as looking into transparent substances, ordinary light radiation is ideal. To see into opaque materials such as the human body, a type of high-energy light, called X rays, is required. To penetrate metals and other very dense objects, even higher energy radiation known as gamma rays is necessary.

Sometimes the task isn't to see into a material but to change it chemically or, in the case of radiation sterilization, to kill bacteria and other biological contaminants. X rays and gamma rays are frequently used for the latter applications, but occasionally other particles may be selected. Some particles may be obtained using simple electrical devices. Others are obtained from radioactive materials that are so unstable that they throw off pieces of themselves. In some cases, particles of such high energy are needed that they must be obtained using sophisticated devices known popularly as atom smashers.

DEMAND GROWS FOR RADIATION INSPECTION

Many modern products have such critical uses that any flaws, including those deep within the object, must be detected. Rotor blades and hubs for helicopters, welds in pipelines that carry toxic or explosive materials, and even artificial hip joints are among the growing list of thousands of products whose interiors must be examined, usually using X-ray and gamma ray radiography. Even the heaviest metal parts, such as entire engines, can be penetrated with high-energy neutrons.

Industrial radiography equipment has been vastly improved in recent years by adding computers to the systems. Equipment is now available that displays radiographic images directly on electronic monitors for continuous, on-line inspection without the need to take or develop photographs. Using a technique called digital radiography, interior views are very clear and may be rotated or enlarged by the operator.

Radiation can also be used to take measurements. When a beam of particles passes through an object, it loses intensity in proportion to the object's thickness. As a result, radiation scanners are frequently used in high-speed production applications to monitor product quality. Sheet metals, rubber, paper, ammunition cases, and many other products are kept to specifications using radiation measurement systems.

Several factors are boosting the sales for radiation-based industrial systems, including reduced system prices, expensive liability suits that follow the failure of critical products, and the need to cut costs by reducing scrap rates.

MEDICAL IMAGING SYSTEMS ARE ALSO BECOMING ESSENTIAL

Closely related to digital radiography for industrial needs are systems designed to make high-quality images of the interior of the human body. Computerized axial tomography (CAT) scanners in particular have found their way into most modern hospitals and will probably be followed by magnetic resonance imaging (MRI) systems that provide images without using penetrating radiation. In addition, newly upgraded X-ray machines and instruments for measuring the flow of low-level radioactive drugs in the body have been developed that have great promise.

The medical imaging industry has suffered from shrinking hospital revenues and reduced medicare payments for expensive, high tech tests. Naturally, medical imaging stocks have fallen drastically in price. However, there is little doubt in my mind that the best of the medical imaging companies will weather the storm because their products save lives.

It is also true that although the unit prices for CAT scanners and MRI machines are high, they are quite cost efficient when kept in use. One $2,000 scan that takes only a few minutes to perform can take the place of many complex tests that could keep a patient in the hospital for a week. In addition, the prices for medical imaging systems are dropping at the same time that capabilities increase.

I'm not surprised that several medical imaging companies have demonstrated increased vitality in recent months and may be emerging from the doldrums. If there ever was an industry in the right position for a turnaround, this is it.

A LOW-COST CONTRIBUTION TO MEDICINE

One of the fastest growing areas in radiation technology involves the sterilization of medical instruments and supplies. Many disposable medical products that are used in great quantities are made of low temperature plastics that can't be sterilized with heat, the traditional method. Until recently a highly toxic gas, ethylene oxide (ETO), and other chemicals were used to sterilize plastic medical products. However, the toxic substances were found to be quite harmful to humans as well as bacteria, even in trace amounts, which caused the government to greatly restrict their use. As a result, sterilization by radiation, usually gamma rays, is becoming the preferred technique for treating many medical products.

Radiation sterilization is now used on over 30 percent of all prepackaged disposable medical devices in the United States, and the percentage is growing steadily. I think the level will probably reach well over 50 percent by 1987. To a large extent, growth is accelerating beyond original expectations because the technique has opened the medical market to many new materials that could not be sterilized by any other method. Radiation sterilization is creating its own growing market at the same time it serves an ever-increasing proportion of existing needs.

FOOD STERILIZATION IS ALSO BECOMING A BIG BUSINESS

Bacteria on foods can also be killed using radiation. Although the technology may appear to be new, in actuality it has been used on a limited basis in 28 countries, including the United States, for several years. As early as 1964, the U.S. Food and Drug Administration approved radiation treatment for potatoes, wheat, and flour to prevent premature sprouting and spoilage. In July 1983, the FDA broadened its approval to include spices and seasonings, and in February 1984, the FDA proposed a rule that will expand the list of approved foods even more by allowing all grains, fruits, and vegetables to be irradiated. The new proposal is expected to gain final approval sometime in 1985. Meats, seafood, and poultry will probably be added to the list by late 1987.

There has been no public rejection of irradiated foods, largely because the process leaves no residual radiation whatever. In addition, foods that have been irradiated have a greatly reduced need for the chemical preservatives that are decidedly unpopular due to the very real possibility that they present long-term health risks.

Of course, food irradiation has the potential of attracting controversy as it becomes more widely used. However, since irradiated foods have a good, long-term track record and offer clear benefits over present preservation techniques, their continued acceptance seems likely.

In many situations, food irradiation makes a very great deal of sense economically, which will undoubtedly be the biggest factor in the technology's success. Not only does irradiation require less than one third the energy used in canning and freezing, but it is also much simpler and less expensive to control. In addition, since workers mustn't be exposed to radiation, the new sterilization systems are highly automated, resulting in sharply reduced labor costs.

Growers and merchants are also waking up to the advantages of irradiated foods. The government's recent ban on the popular fumigant ethylene dibromide (EDB) is providing a powerful incentive among citrus growers to consider irradiation. Since irradiated foods have an extended shelf life and a greatly decreased need for refrigeration, grocery stores are giving them an enthusiastic welcome.

All in all, food preservation by irradiation is a well tested and economically attractive technology that promises rapid growth over the next few years.

CREATING NEW PRODUCTS WITH RADIATION

Several types of very commonly used plastics start out as syrupy liquids composed of single chains of molecules. After the addition of another material, the chains begin to form chemical bonds with each other in a process called cross-linkage, causing the liquid to become a solid.

Radiation can be used instead of chemicals to cause liquid plastics to form cross-links and turn into solids. Because radiation can probe deeply into most materials, it is a much more effective agent than chemicals for curing plastics that have been squeezed deep into crevices. A very common example is plastic-impregnated wood used in knife handles and parquet floors. Liquid plastic is placed under great pressure and forced throughout the wood, then turned into a solid with a dose of radiation. The result is a very popular composite material with the best properties of both components. Many other radiation-treated products are found in the rubber, metal, and electronics industries. In fact, radiation is now essential to the production of several new types of semiconductors and semiconductor packages.

The chemical industry is also developing greater uses for radiation. Several high-margin products can be made in a purer form or made more quickly when radiation is used instead of traditional methods. Although radiation chemistry is still a young science, it is developing rapidly.

DEFUSING A TICKING BOMB

It is frightening to consider the amount of nuclear waste that has been allowed to build up since the beginning of the atomic age. Almost all the radioactive by-products generated by the nuclear power and defense industries are still in existence in unsafe, "temporary" storage. Over the next several years, millions of dollars will be spent creating permanent treatment methods and repositories for radioactive waste products.

A great deal of the money spent on nuclear waste will go to the environmental tech industry. (See Section 5, earlier in this book.) However, there will be a great need for precision radiation-monitoring instruments and specialized equipment for handling radioactive materials that the radiation tech industry will supply. Orders for the latter products are already being made and will surely accelerate in the future.

INVESTMENT OUTLOOK

As can be seen, the radiation technology industry has a bright side not yet noticed by most investors. Many companies in the industry display enormous vitality and have promising futures. I expect this field will prosper and attract considerable attention before long. Hopefully, by then, many investors will have taken solid positions for long-term gains.

Radiation Technology Companies

Company	Symbol	Exchange
ADAC Laboratories	ADAC	OTC

Comments: ADAC has been a leading medical imaging company that also offers products for radiation therapy and nuclear medicine. The company's computerized systems are used in more than 1,500 medical centers. The company grew rapidly for several years but lost money in 1984, as did many medical imaging firms. New management and products look encouraging.

Diasonics	DNIC	OTC

Comments: Diasonics is a top medical imaging company that offers ultrasound systems in addition to mobile X-ray and MRI equipment. After suffering severe losses in 1983 and early 1984, the company appears to be turning around. Diasonics has potential and is definitely a company to watch.

Elscint	ELSTF	OTC

Comments: Elscint is a pioneer in medical imaging systems and remains a leader in the field. The company makes digital X-ray equipment, CAT scanners, MRI systems, ultrasound devices, and a variety of nuclear medicine products. Well worth following.

Fonar	FONR	OTC

Comments: Fonar recently received FDA approval to market its MRI scanners. The company differs from most other MRI suppliers in that it uses permanent magnets that require no outside power. The company hopes to have several machines installed by midsummer 1985.

Imatron	IMAT	OTC

Comments: Imatron makes high-speed CAT scanners. As of Spring 1985, five of the company's machines were being used for research in the United States.

IRT	IX	ASE

Comments: IRT has a broad representation in the radiation tech industry, with contributions to hazardous material control, radiation monitoring, automated inspection, industrial radiography, and electronic system survivability. The company is one of the few companies in the United States doing extensive work on computer and communication systems that could be exposed to electromagnetic pulse (EMP) and intense radiation following a nuclear explosion. (See related information in Chapter 5, this section, on security.)

Company	Symbol	Exchange
Isomedix	ISMX	OTC

Comments: Isomedix is the leading medical sterilization company in the world. The company has a network of irradiation centers in the United States, Canada, and Puerto Rico. In addition to its medical irradiation services, Isomedix is quietly positioning itself to conduct food sterilization. The company's irradiation centers have been designed to accommodate the necessary changes to permit the new applications, which should contribute substantially to revenues in the future. No other company appears to be in a better position than Isomedix to capitalize on the growing sterilization business.

Matrix	MAX	ASE

Comments: Matrix is a company whose primary business is recording computer-generated images. The company is a leader in its field. Matrix makes its specialized components for several medical imaging devices and represents a nice way to play the industry. The company showed excellent results in fiscal 1984, which was a tough year for front-line medical imaging companies. Not a pureplay but definitely a company to watch.

NMR of America	NMRR	OTC

Comments: NMR supplies magnetic resonance (MRI) machines primarily for outpatient imaging by private physicians and medical centers. This small company is a pureplay in a field that may grow as the high costs of machines decline.

Nuclear Support Services	NSSI	OTC

Comments: Although the nuclear power industry doesn't appear to have a very promising future, existing plants must still be serviced. Nuclear Support is one of the few companies that offer such services. These will be even more necessary as our nuclear plants grow older. The company should find its business increasing for several years.

Radiation Technology	RTCH	OTC

Comments: Radiation Technology is an irradiation company with an emphasis on food sterilization. The company also uses radiation to cure plastics for several applications. Radiation Tech is a very small company that has the potential to grow rapidly. The company is well worth watching.

Tech Ops	TO	ASE

Comments: This company is a leading supplier of industrial radiographic inspection and imaging systems used primarily to examine welds and castings in jet engines, pressure vessels, and other high-stress products. In addition, the company offers radiation-monitoring services to nuclear power and other industries that work with radioactive materials.

Other public companies with substantial involvements in the radiation technology industry include: Nuclear Data (NDI, ASE, radiation instrumentation), Nuclear Pharmacy (NURX, OTC, radiopharmaceuticals), and Scintrex (SCT, Toronto, radiation survey instruments).

CHAPTER 40

Automation Reaches the Printing Industry

In this section's introduction I mentioned that the archaic nature of much of the technology used in the printing industry stands in sharp contrast to the modern machines I see during my high tech investigations. The problem isn't confined to the presses. It involves the whole production process, especially the front end, which is the most labor intensive.

Incredibly, much work is still done on typewriters or standalone word processors that have no direct connection with the presses. Instead of doing a job once, then having it roll off a press, in most shops it must be handled many times. Text is typeset, then pasted up, strip by strip and block by block. Illustrations are enlarged or reduced, photographed, then pasted up with the text, after which the whole thing is again photographed. The photographs are then developed and a printing plate is created.

Clearly, the printing and publishing industry is long overdue for modernization and represents a rich market for those companies that produce the necessary equipment.

SOLUTIONS ARE BECOMING AVAILABLE

Despite my initial observations, efficient, computer-based publishing and printing equipment has been developed. The new systems are beginning to sell very well to industry leaders who are finding it essential to their survival. Although many members of the industry have already started to modernize, the great majority has yet to do so. As a result, I expect to see substantial growth in this industry over the next few years.

The most cost effective of the new printing automation systems are computer-based text editing and layout devices. Such devices greatly simplify the production of most publications by combining many functions in one system. Typewriters, word processors, typesetting equipment, layout boards, and even cameras have been replaced by the new systems that do everything electronically. The most advanced equipment of this type shares information with a central computer and with other workstations in the network. Text may be entered, edited, merged with other material, and finally laid out in final form for the press. As we saw in the section on image digitizers (Chapter 20), even graphics can now be brought into the system.

Computerized systems are also available that automatically prepare the photographic masks used to make printing plates. The first such systems, which used stored "slides" of each individual letter, have been replaced with high-speed digital devices that draw letters on the photomasks using high-speed lasers. The newest photocomposition systems skip the photomask steps altogether and "burn" a plate directly with the laser.

Lasers are also being used in a new type of printing system based on photocopy technology. Instead of using a laser to make the printing plate that is used on a press, the laser writes each page on an electrically sensitized drum similar to those used in copiers. The drum then picks up toner (about the same as ink) in the desired places and prints the complete image immediately. Because laser printers are incredibly fast, one day soon many publications will only be printed as they are needed. This book, for example, could be printed and bound in just a few minutes with the new laser-based publishing systems.

EVEN THE PRESSES HAVE BEEN AUTOMATED

Conventional though thoroughly modern printing presses have also been developed that are far more efficient than their "steam engine" predecessors. The new presses are as advanced as computerized machine tools found in automated manufacturing installations and often employ robotic and process control technology. Indeed, the major advances in presses occurred after process control experts were brought into the picture and instructed to treat printing as if it were a small-scale automation problem. As a result, the latest printing systems more closely resemble fully automated paper or sheet metal mills than they do presses. The increased productivity they bring to the printing industry is making their adoption mandatory.

The last step in the modern publishing and printing equipment industry is to finish linking the new systems together into fully integrated units. In addition to those procedures directly related to producing finished copy, many business functions such as automated billing are now being tied into the systems. With some equipment it is possible to lay out an ad, calculate its cost, and bill the customer, all automatically.

INDUSTRY TRENDS

The market for high tech printing equipment is actually much larger than may first appear. In addition to the traditional printing and publishing companies, many businesses are proving to be ready customers. What plastic pipe did for do-it-yourself plumbers, high tech printing equipment is doing for many industries with brochures and catalogs to print. Prices are coming down, with the result that much of the equipment that was formerly affordable only by the big publishing and printing firms is now within the range of many small businesses and print shops.

Within the next few years, complex text editing and photocomposition systems, along with laser printers, will take their place with word processors and microcomputers in the efficient office. An end to the expansion of the market for high tech publishing and printing systems is not yet in sight.

Printing Automation Companies

Company	Symbol	Exchange
Compugraphic	CPU	NYSE

Comments: Compugraphic is the leading supplier of computerized typesetting systems. Products are sold to newspapers, magazines, and large businesses with in-house publication needs. The company has an excellent reputation for quality products and service.

Company	Symbol	Exchange
Harris Graphics	HG	NYSE

Comments: Harris Graphics acquired its independence from Harris Corporation in April 1983. The company is the world's largest manufacturer of computerized web offset printing presses, which are used to create full-color magazines, catalogs, and similar products. Web presses print on continuous rolls rather than individual sheets and tend to be very large, complex machines. Harris designs its presses as if they were small, self-contained factories. This is a top company well worth watching.

Company	Symbol	Exchange
Penta Systems International	PSLI	OTC

Comments: Penta Systems competes in the same markets as Compugraphic but is placing special emphasis on the growing demand for in-house publication by businesses. The company is merging its systems with office automation equipment, which is undoubtedly the correct strategy. It remains to be seen if this small company will be successful. If so, its fortunes will be assured. I recommend you follow Penta Systems' progress.

Company	Symbol	Exchange
System Integrators	SINT	OTC

Comments: System Integrators makes what could best be described as the world's largest word processors and photocomposition systems. The systems were designed primarily for newspapers and magazines and consist of a central computer and many workstations. Individual articles, ads, graphics, and other information is fed into the system, which is then used to create the finished product electronically. Billing and other business functions are also performed on the system. Definitely a company to watch.

Other public companies with substantial involvements in this industry include: Bedford Computer (BEDF, OTC, photocomposition and graphics equipment) and Information International (IINT, OTC, prepress production equipment).

CHAPTER 41

After the Shakeout, High Tech Stores Will Prosper

As this is written, a shakeout is occurring in the retail microcomputer industry. Consolidation is progressing rapidly as successful chains squeeze out the independents whenever they come in contact. Smaller chains are running up red ink, quarter after quarter. Many will not survive into 1986.

In the midst of the carnage, strong companies are emerging that will find themselves with a clear field in which to do business. When the smoke clears, good profits will begin to return to selected specialty stores with unique products and markets. By then, so much negative press coverage will have followed the industry that many investors will not think to look for the attractive positions that will exist.

In addition to the shakeout, other trends are developing that have important implications for investors. It is apparent that although low-end microcomputers and microcomputer products have become commodity items to be found in large department stores and other mass merchandisers, middle- to high-end microcomputer products remain firmly with the specialty stores. The most successful of such stores cater to business executives and professionals who will pay top dollar for expert assistance in selecting their computer systems.

At some point, however, customers will become sufficiently

comfortable with high tech business systems that they will not need the help of experts to make selections or learn how to use them. At that point, even sophisticated equipment will find its way into less specialized discount stores. During the first phase of that development, a strong market will exist for retail chains that rent space in department stores. Such semispecialty stores within stores will offer the best of both worlds, reasonably well informed salespeople and moderate prices.

VARIETY IS IMPORTANT

Another trend that is emerging is the growing strength of stores that handle a variety of high tech business products in addition to computers. We are beginning to see the more successful stores offer copiers, electronic typewriters, feature phones, PBX systems, and specialized business publications, along with computers. Customers obviously prefer one-stop business shopping over running around to separate stores and are more likely to frequent retailers that have such a variety of products. By having many different items to sell, stores have more ways to cash in on the good will and rapport they build, patiently and expensively, with clients.

REACHING FOR MARKET SHARE

Expansion is a major objective for retail business systems chains that obviously feel that "getting the mostest the fastest" is a major key to capturing market share and ultimate success. However, there are two schools of thought regarding the best way to gain territory.

One method involves outright ownership of each store, even though the enormous capital requirements of such an operation make it extremely difficult and dangerous to add stores when profits are thin. The proponents of outright ownership argue that losing some expansion capability is a fair exchange for having absolute control over each store and perhaps a better chance to maximize profits over the long term. It is becoming increasingly apparent, however, that many proponents of total ownership will not survive to enjoy the long term.

Other chains use franchises to avoid the huge capital requirements of outright ownership and are able to continue expanding even when profits are thin. They are somewhat less concerned with having total control than are the proponents of outright ownership. Managers of franchise chains are convinced that profits are best maximized by having a large number of stores in the network. Franchise operations have proven themselves in fast-food and other industries and may do so again with business systems stores.

WATCH HOW THE TRENDS DEVELOP

It is a certainty that high tech business products will continue to sell well, although fluctuations due to the economic climate will have an effect. It is also clear that people do not like purchasing delicate products through the mail. As a result, local sources will get the lion's share of the business. The trends that are emerging for such sources are worth following by astute investors in search of opportunities not seen by others.

Business Products Stores

Company	Symbol	Exchange
Businessland	BUSL	OTC

Comments: Businessland is a national chain that owns its own stores. The company is quite innovative in its approach, which distinguishes it from many of its competitors. Besides serving walk-in customers, each store in the chain is a regional sales office for personnel who market products directly to large corporations. In addition, Businessland offers a wide range of products, from microcomputers to copiers. Definitely a company to watch.

Company	Symbol	Exchange
Computer Depot	CDPT	OTC

Comments: Computer Depot is unique in the microcomputer retail industry because it rents space in major department stores, giving the company instant credibility and a large flow of potential customers. The company may be in the right position for the next phase of computer sales, which is a move from specialty to general purpose stores. Of course, Computer Depot may not necessarily be the company that ultimately benefits from the change. However, you should be certain to investigate this firm's progress before making an investment in this industry.

Company	Symbol	Exchange
Entre Computer Centers	ETRE	OTC

Comments: Entre is one of the largest and fastest growing franchisers of retail computer stores in the United States. The company has clearly targeted its products and sales efforts for the profitable business and professional microcomputer market. The company's aggressive expansion program and its well-defined marketing strategy may prove to be the winning combination.

There are several other public companies in this industry I urge you to review prior to taking a position. They include: The Computer Factory (CFA, ASE), ComputerCraft (CRFT, OTC), Computers For Less (CFLI, OTC), The Computer Store (TCSI, OTC), and Inacomp (INAC, OTC).

CHAPTER 42

America Becomes Security Conscious

The security systems that people notice most often are the antishoplifting systems found in stores, and burglar alarm systems seen in homes and businesses. However, the visible part of the security industry is just the tip of the iceberg. Unnoticed by the general public, very sophisticated personnel ID and verification systems have been developed, which represent excellent long-term investments. In addition, the new field of computer and communication security offers many promising opportunities.

PERSONNEL ID AND VERIFICATION

One of the biggest problems facing government and commercial organizations is restricting entry to buildings and electronic systems to selected personnel. In such applications, fast, reliable personnel verification systems are needed in hugh quantities and are finally becoming available.

In addition to verification systems, which usually rely on

passes and confidential query information, a significant market exists for true identification systems that can pick one person out of millions using fingerprints or other physical characteristics. Incredibly, such identification systems are now available at affordable prices and are being installed in high-security situations and by a growing number of police agencies.

Most of the latest personnel verification systems rely on special cards, tags, and badges that differ primarily in the amount of information they carry and the way that information is stored. Plastic cards with magnetic strips are the most commonly used, primarily for consumer credit applications. Because magnetic strip technology is well established and inexpensive, it is well suited to low-security situations, where it will become even more widely used in the future.

Unfortunately, magnetic cards are unable to hold more than a few pieces of information and, therefore, can do little more than provide basic identification. In many cases, the information on the magnetic strip can easily be decoded and altered. As a result, new "smart" cards and tags have been introduced that contain embedded semiconductors with large data storage capacities and specialized security functions.

Smart cards and tags are easily carried and inserted into special decoder/interrogators that are often linked to computers. Because smart cards can be updated each time they are used, they are ideal for hospital patients, soldiers, and others who must carry their records with them. Smart tags are also well suited for applications such as computer security, where audit trails for file access are desirable.

A technology midway between magnetic memory cards and smart tags are plastic cards with information recorded on the surface by a laser. Cards with laser memory resemble ordinary consumer credit cards but hold enormous quantities of information, even more than smart tags. Unlike smart tags, however, the information on a laser memory card cannot be erased or, for now, easily updated with use. At present, most ID applications either require very little data storage or they require quite a bit of information, which must be changed frequently. As a result, cards that use the new laser technology have yet to find a large market.

Suppliers of personnel security products involving plastic

cards of whatever type tend to make large initial sales of embossers, encoders, and other equipment. Sales are followed by orders for card readers, verifiers, and related devices, which are often needed in large quantities as well. Of course, card blanks themselves will provide companies in this industry with continuing revenue streams for years.

The most futuristic of the new personnel security systems uses biometrics. This system analyzes an individual's physical and biological characteristics and compares them to those previously measured and recorded. If the characteristics of the person using the system match those in memory, positive identification is made.

Unfortunately, many characteristics change from use to use. Consequently, a system that is sufficiently discriminating to reject unauthorized individuals may also refuse to verify many legitimate users. An excellent example of this problem involved a top-rated voice verification system used to control entry to nuclear power plants that was tested extensively. The system passed all the preliminary tests and was close to adoption, but during a simulated emergency, the system created what would have been a disaster. The emergency response personnel hadn't been told the exercise was just a drill. Every one of them was rejected by the system due to the effects of tension on their voices.

For one reason or another, such characteristics as retinal patterns in the eye, hand dimensions, signatures, and voice patterns have not been reliable as verifiers in personnel security systems, although advances in these areas are expected. However, systems using fingerprints for verification have proved themselves to be entirely satisfactory and are the basis for most present-day systems. Besides being unique, fingerprints are easily measured by modern computer-based scanners, and fingerprints don't change with stress, mood, or (for the most part) age. Of course, law enforcement agencies are delighted because tens of millions of fingerprints are already on file.

COMPUTER AND COMMUNICATION SECURITY

The vast majority of computers and communication systems in the United States have little or no effective security. As unbelievable as it may seem, much of the technology on which our society relies is unprotected to such a deplorable extent that severe economic and strategic consequences could easily result. Fortunately, major steps are being taken to correct the situation before it is too late. The process will take many years and will bring millions of dollars into the security industry.

The problem isn't confined to computer "hackers" and other unauthorized persons and agencies who gain access to electronic records through phone links. A far bigger problem is intentional and accidental misuse of information by people who do have authorized access to computers and networks but who are not adequately prevented from using restricted parts of the systems. Experience has clearly demonstrated that passwords and ID codes are woefully inadequate to prevent unauthorized access. Incredibly, too few of our systems even have password protection.

Fortunately, effective callback systems and high-security software is now available that greatly improve computer security. Considering how few computer systems have adequate safeguards, demand for the better protective technology will almost certainly exceed supply for several years. Many small computer and communication security companies that have quietly developed top products are well positioned to make a very great deal of money.

A less publicized problem with computers is their vulnerability to electronic eavesdropping. Unless computers are adequately shielded, they emit telltale signals about the information they are processing that can be picked up from a considerable distance. The job of protecting computers from being monitored as they work has barely started.

Perhaps an even bigger problem than the lack of computer security is our wide-open telecommunication system. Billions of dollars of electronic fund transfers and strategic information of every type is literally handed to anyone with a few thousand dollars worth of receiving equipment. Although effective encryption systems and secure modems are now available, they are so new that well over 95 percent of the market has yet to install them.

Of less concern to individuals but of great importance to the government and large telecommunication companies is the vulnerability of our microprocessor-based electronic systems to the paralyzing effects of even small nuclear explosions. Such explosions emit powerful electromagnetic pulses (EMP) that scramble integrated circuits. As a result, a relatively small number of low-yield nuclear airbursts could stop the nation dead in its tracks while doing little physical damage. The job of "hardening" vital electronics against EMP is quietly underway but will take years to complete.

No discussion of communication security would be complete without mentioning the growing problem of industrial espionage. Restricted information may be obtained in a variety of ways, but especially vulnerable are discussions between people within an organization where they feel secure. Unfortunately, as many firms have learned, security can no longer be taken for granted. In our highly competitive world, where millions of dollars ride on single products or decisions, thousands of firms are finding it necessary to spend large sums on electronic countermeasures against phone taps and hidden radio transmitters. The industrial espionage problem, and the business it creates for the security industry, will undoubtedly continue to grow.

INVESTMENT OUTLOOK

Although the market for high tech security systems is expanding rapidly, there are relatively few companies in the field. Even better news is that most of the leading suppliers are publicly held. Lastly, for all the potential this industry offers, it has yet to gain widespread attention.

High Tech Security Companies

Company	Symbol	Exchange
Data Card	DATC	OTC

Comments: Data Card is a leading supplier of magnetic strip plastic cards, card makers, readers, and other related devices. The company's products are found primarily in consumer credit applications but have been adapted for purely identification purposes by many customers. The company is currently investigating new cards with embedded microprocessors and has an agreement to market smart keys made by Datakey (see below). Good long-term outlook, but is very sensitive to the economy due to its consumer credit involvement.

Company	Symbol	Exchange
Datakey	DKEY	OTC

Comments: Datakey is a small company that makes smart plastic badges, keys, and passes that contain microprocessors. The company's products are presently used to control access to critical medical and computer equipment and are being tested by the U.S. Army as possible replacements for dog tags. Datakey also supplies devices to encode and decode its products. Smart tags have yet to become popular in the United States, but if they do Datakey may prosper.

Fingermatrix	FINX	OTC

Comments: Fingermatrix is the present front-runner in the biometric industry. The company offers fingerprint verification systems that are affordable and compatible with FBI matching procedures. Systems are in use primarily in area security applications, although various law enforcement and government agencies are also evaluating them. This company appears to have excellent potential.

IRT	IX	ASE

Comments: IRT is a leader is EMP survivability. Please see listing under radiation technology, Chapter 2, this section.

Spectrum Control	SPEC	OTC

Comments: Spectrum is a leader in the field of electronic emission control and shielding. The company supplies products that not only prevent electronic devices from emitting signals, but also from being adversely affected by EMP and other interference. Because Spectrum Control is in on both ends of computer and communications security its prospects should be especially good.

Systematics General	SYSG	OTC

Comments: This company supplies computer and communications equipment primarily to agencies of the U.S. government and the military or to other firms that supply such markets. The company's products have been engineered to prevent them from emitting signals that could be intercepted by unauthorized persons.

Technical Communications	TCCO	OTC

Comments: TCC is a very broadly based communications security company. TCC offers scramblers, encryption and ciphering systems, and many other high-security products for voice and data telecommunications. Some products are portable. Should be followed.

Company	Symbol	Exchange
Verdix	VRDX	OTC

Comments: Verdix is one of the few companies that is developing a computer system from the ground up that is designed to offer a very high level of security. The new machine should be ready in late 1985. In addition, the company is focusing attention on the new computer language, Ada, that is being adopted by the U.S. military. This is definitely a company to watch.

Other public companies with substantial involvements in those subsectors of the security industry covered in this section include: CGA Computer (CGAC, OTC, security software), Drexler Technology (DRXR, OTC, laser strip cards), International Mobile Machine (IMMC, OTC, computer and communication security), and National Business Systems (NBSIF, OTC, magnetic strip cards).

Part II

Investment Vehicles and Strategies

Section 11

Choosing the Type of Investment That's Best for You

CHAPTER 43

Many Paths to Profits

High tech investors are fortunate in having a wealth of attractive opportunities from which to choose. No matter what your personal characteristics or your investment objectives may be, there is an investment that is suited to you. The most difficult part of the selection process is to honestly assess your own position.

Aggressive investors who have the time and the interest to follow their investments closely may find high tech new issues a very attractive means to profits. Analyzing new issue prospectuses will not only make it possible to understand and evaluate individual companies but it will also acquaint you with whole industries and technologies. Prospectuses tend to be superbly written, although not from a stylistic standpoint, and are very educational. I've never met an experienced new-issue investor who hadn't become very knowledgeable about the areas he or she preferred.

Individuals who are both aggressive and short-term oriented should find the new technology index options perfect for them. One index option exists just for the computer industry and another reflects a broader selection of technology companies. In both cases prices can change very quickly and all bets are called within three months.

At the other end of the spectrum, but no less attractive, are high tech mutual funds that require little direct involvement after the initial selection has been made. For busy people or those who don't like the tension that often goes with hands-on investing, mutual funds may be ideal. There are many to choose from and several have excellent track records.

At the midpoint between mutual funds and new issues are high tech growth stocks that require active but usually not daily involvement. Growth stocks are probably the most popular of the many high tech investments. They offer outstanding appreciation without as much volatility as is common with new issues.

Off by themselves are venture capital companies that operate much like closed-end mutual funds. Although they specialize in start-ups, they are much less volatile than high tech new issues. Investors have yet to take much notice of venture capital companies, but that situation should change in the future.

Whatever your choice of investments may be, it is important to get started if the markets are cooperative. When it comes to amassing wealth by compounding profits, it is much more important to increase the number of successful trials than it is to wait until you have a large nest egg with which to start.

CHAPTER 44

Taming High Tech New Issues: You Can Turn Roulette into a Science

High tech new issues can be very profitable when they are selected and managed skillfully. In fact, with only a little effort, from mid-1983 through late 1984, many careful investors made excellent gains in the new issues market overall and many times their average returns on several individual offerings, including Applied Biosystems (64 percent), ARGO Systems (78 percent), Ceradyne (43 percent), Daisy Systems (54 percent), Metromail (45 percent), Micron Tech (89 percent), and Telco Systems (63 percent). Moreover, they chalked up their profits in spite of long periods of poor market conditions, when thousands of conservative investors were losing money. Of course, many high tech new issues also performed very poorly during the period in question.

Please notice that I didn't include Digital Switch, Convergent Tech, Flow Systems, or any of the other big-name new issues in the previous list. I'm not interested in occasional lucky shots, however impressive they may be. Instead, I wish to show how cautious, conscientious investors can make attractive profits year after year by carefully selecting the most promising high tech new issues that come available regularly.

The volatile price action of many new issues frightens most people, and rightly so, but the volatility can be the investment's greatest advantage. Because price movements tend to be sharp and decisive, trends are often revealed early, which makes it possible to cut losses quickly and identify winners. If you are careful to start out with only the best of the proposed offerings, the volatility of high tech new issues can be exploited to fine-tune a portfolio for outstanding returns. With high tech new issues, as in many other investments, what appears to the majority of people to be a great liability can be used by others to their advantage.

GETTING IN ON THE GROUND FLOOR

All too often, investors interested in new issues don't learn about them until the initial public offering (IPO) is long over and the company has attracted the attention of the financial media. By that time, the lion's share of the profits have often been made. However, early birds can learn about attractive offerings well ahead of the IPO through *The Investment Dealers Digest*, Standard & Poor's *New and Emerging Situations*, Value Line's *New Issues Service*, and Barrington Research's *High Tech Investor*. The latter three publications review many top offerings and make recommendations. (See the list at the end of this chapter.)

Once you've learned about an upcoming IPO and have made some tentative selections, you should contact the underwriters and request the preliminary prospectuses you will need to make your final choices. If time is short, you should phone the underwriters and ask for the prospectus department. Be prepared for someone who will probably talk like a machine gun and have no time whatever for chitchat. But the prospectuses are worth getting, whatever the irritation. They are the best reports investors can get anywhere on Wall Street because the SEC demands that each contain a full disclosure of all the facts relating to the probable success or failure of the company. Prospectuses transform the new issues market from a high-risk game of chance into an effective money-making investment.

SELECTING THE CREAM OF THE CROP

Much of the risk with high tech new issues is reduced and the chance for profits increased if selections are limited to those launched by top technology underwriters, including Alex. Brown & Sons; Hambrecht & Quist; Robertson, Colman & Stephens; and L.F. Rothschild, Unterberg, Towbin. Their addresses, as well as those of other respected underwriters, are listed in Part III.

The leading underwriters commit themselves to purchase the shares of the firms they sponsor, so they are careful to select only high-quality issues that they can sell to the public. Although the research done by the leading underwriters is certainly no guarantee that their new issues will succeed, it is nevertheless a screening process of considerable value to investors.

When a prospectus arrives, one of the first sections you should review is the brief description of the company's business and products, usually found on page 3. If you find the firm is still in the development stage, I suggest you hold the prospectus at arm's length, as if it were a dead mouse, and drop it in the trash. It's hard enough for a young company to prosper even when it has top products and good marketing going for it. Without those advantages, the company's chances are very small. As we've seen with dozens of promising development-stage companies, great ideas and noble efforts just aren't enough for success. Fortunately, there are many companies each year that go public after their great ideas and noble efforts have become highly marketable products. Those are the new issues to buy.

By far the best strategy is to take positions in small but solid companies that have established products, aggressive marketing plans, and who are firmly positioned in rapidly growing industries. Buying such issues gives you two rides for one ticket—you get the growth of the company plus the growth of the company's industry. Such industry leverage often results in staggering appreciation, as it did with Genentech in biotechnology, Digital Switch in data communications, and many other well-known new issues.

While you are scanning the summary, be certain to see if there is a trend of steadily increasing sales and earnings. Be suspicious of a company that was losing money, then suddenly hit the jackpot just before its IPO. It might be a coincidence, but even so one or two good quarters isn't much of a track record.

You'll also find a section called "Principal and Selling Shareholders" in the prospectus. It is a list of the company's backers and officers. You should read this section to see if any of them are trying to sell very much of their stock. Of course, many founders and original investors are likely to cash in some of their holdings as compensation for being on short rations during the company's formative years, but if any of them are selling more than 35 percent of their shares, you can expect that they have little faith in the future of the enterprise and neither should you. On the other hand, if they are all hoarding their shares, maybe you should try especially hard to get some.

The section "Use of Proceeds" is also important. The company should need most of the money from its IPO for expansion rather than paying off debt. Although some debt payment is OK, you should expect to find evidence of a strong, healthy company that needs the money to maintain its leadership position, not to keep the wolf away from the door. Fortunately, the prospectus makes it easy to tell the difference.

Other sections of the prospectus deal with such topics as competition, customers, and risks. They are usually quite easy to evaluate, but here are a few items to watch for. Although it's a good sign if the company has a long list of reputable customers and lots of repeat business, no single account should be so important that its loss could cripple the firm. As to competition, you should expect it, but be sure the company isn't fighting an uphill battle with an industry giant. Lastly, don't necessarily be carried away by pessimism after reading the section on risks, often titled "Certain Factors." The SEC's full-disclosure policy is admirable, but sometimes the sheer number of risks that are discussed can paint an overly negative picture. Evaluate each risk on its own merits.

Reading through each prospectus may be a bit slow at first. Fortunately, the documents follow a standard format and use the same key phrases to reveal important facts about companies, so after you've been through a few you will become proficient at quickly spotting the most promising offerings. You will soon join the ranks of many successful investors who know that prospectuses make new issues among the easiest investments on Wall Street to critically evaluate.

HOW TO BUY NEW ISSUES

A high tech new issue can be difficult to get at the initial offering price, especially if it's a "hot" deal in a strong market. When there's not enough stock to go around, it's often allocated by the firms in the selling group to their top-performing brokers, who in turn favor their biggest clients. If you are a good customer of such a broker, your chances of getting your stock are good. If not, ask your broker to request that the managing underwriter obtain shares for you. If that doesn't work, you can contact a local office of the underwriter or ask the officers of the company to go to bat for you. If your time is valuable, however, I suggest you pay a bit more and simply buy your shares in the aftermarket that is established right after the opening. A new issue that is in such demand that it can't be obtained at the IPO should continue to appreciate and make the small premium you paid insignificant.

If you miss catching an attractive new issue in the early aftermarket and feel it has moved up too much in price, I suggest you keep your eye on the situation, as you may be handed an excellent buying opportunity later. Unexpected bad news about a highly publicized young company can often send the price of its stock tumbling because so many of its initial investors were simply joyriding. If the bad news does not seem to affect the long-range potential of the company, you'll have a bargain.

The aftermarket is almost always the best place to shop during sideways or down markets. You should take your time about it, perhaps as much as several months, because new issues, including those of very high quality, sometimes lose as much as 50 percent of their value during poor market conditions when investors flock to blue chips for safety. Although the loss is predictable, casual investors are horrified by it. As always, more knowledgeable investors find opportunity where others see disaster. They make their selections carefully, then wait patiently for the expected price erosion to turn them into bargains.

DIVERSIFY, THEN CUT YOUR LOSSES

No matter how carefully you select new issues, you will certainly make mistakes. The damage such mistakes can do to your overall investment success can be held to acceptable levels by diversifying your new issues portfolio, then ruthlessly cutting losses. Although there are no formulas for cutting losses, experience has proved that they can be held well below average by conscientious investors. This is especially true if you start out with the cream of the new issues crop and if you make most of your purchases in the mature aftermarket.

When hot deals are purchased and sold in bull markets, one's timing can be critical, because price volatility is far higher than is true is quieter periods. Although the techniques and objectives are the same in both markets, you have less time to think and react when the pace quickens. However, you can give yourself more breathing room and boost your profits as well if you watch for postcorrection bargains as previously discussed and take a longer range view toward new issue investing.

In buoyant markets, a popular new issue with great products but no earnings may appreciate to ridiculous levels. Without earnings to provide an objective measure of performance, investors sometimes let their optimism, and their greed, get in the way of good sense. In such cases, the last thing you want to see the company do is make a profit because everyone will then be able to set a reasonable price! When management begins to boast about "finally moving into the black next quarter," it is time to sell. Better yet, stay away from these deals altogether.

MAKING IT WORK FOR YOU

Because high tech new issues attract many people whose approach is more suitable to Las Vegas than it is to Wall Street, the field is too often ignored by serious investors. Yet there are few areas that offer the potential of high tech new issues, provided this field is approached with care and skill. Investors that do so are finding the field is wide open and very profitable. Each year, even during bad markets, high tech new issues appear that appreciate substantially. As we've seen, finding them is a great deal easier than most investors expect but you do have to do your homework. Although you won't need any really big winners to be successful, sooner or later you're likely to find one in your carefully selected portfolio.

News about New Issues

Emerging & Special Situations
Standard & Poor's
25 Broadway
New York, N.Y. 10004
Monthly, $130 per year

High Tech Investor
Barrington Research Associates
P.O. Box 860
Barrington, Ill. 60010
Monthly, $95 per year

Investment Dealers Digest
150 Broadway
New York, N.Y. 10038
Weekly, $195 per year

Technology Stock Monitor
Barrington Research Associates
P.O. Box 860
Barrington, Ill. 60010
Monthly, $180 per year

Value Line New Issues Service
711 Third Avenue
New York, N.Y. 10017
Twice Monthly, $300 per year

CHAPTER 45

High Tech Growth Stocks: They Can Make You Wealthy

High tech growth stocks have probably made more people more money than any other investment on Wall Street. They combine the potential for price appreciation common to new issues with some of the safety of larger firms. The blend of characteristics seems ideally suited to today's busy investors who are able to watch their investments closely but who can't devote full time to the job.

Growth stocks are also well suited in terms of the relatively short amount of time they usually take to succeed or fail. The quick maturation time is very important to many of today's investors, who usually must start out with small amounts of money and who know that it will make many sequential investments to build substantial wealth. Such investors know they can get more successful trades in a given amount of time with growth stocks than they will with blue chips and are willing to accept higher risks and more work to do it.

Individual investors are not the only ones to climb on the growth-stock bandwagon. Institutional investors are also making more use of the smaller issues. They have partially solved the problem of being too big to participate in the smaller stocks by allocating manageable blocks of funds to outside professionals to handle according to their individual areas of expertise. The institutional move to growth stocks is quite bullish, but it does increase price volatility when many of them happen to move in unison.

IDENTIFYING AND SELECTING PROMISING GROWTH STOCKS

A few years ago I developed a selection checklist for high tech growth stocks that has a good track record for identifying winners and keeping losers out of a portfolio. The checklist is also a good description of technology growth stocks and is worth examining.

Size

The companies with the greatest combination of high growth with reasonable safety have sales under $50 million a year. The farther sales get below that level, the easier it is for a company to double and triple them. However, very low annual sales of $1 million or $2 million often indicate the company is too small and may still be a long way from reaching its stride. The objective is to catch a company that has proved itself but is a long way from its ultimate size.

Growth

The company's annual growth should already have reached 25 percent. In the high tech fields with the most promise, much less than a 25 percent growth rate for a small company can indicate something is amiss. If the firm is really positioned as well as it should be, the growth rate will climb well above 25 percent during its prime years.

Industry Leverage

As I mentioned in Chapter 44, this Section, on high tech new issues, industry leverage is extremely important. The growth of the company's industry acts as a multiplier on the growth of the company itself. The result can be breathtaking. In fact, a growth company in a growing industry is one of the world's top combinations for profits. You can think you won the lottery with the right combination of company and industry.

Leadership

The company should definitely hold a leadership position in its industry. It is far easier for a company to stay on top than it is to climb up and push another firm aside. Look for an existing situation to continue. Never bet on underdogs. In many small, developing industries there may be room for only one top firm. In such situations, everyone else will be in a very poor second place.

Products

Often a technology company will make a lot of money simply by being "the firstest with the mostest" in a rapidly expanding market. However, if its products aren't patented or are not sufficiently sophisticated that others can't make them, its profits may fade quickly.

Products that improve productivity have the greatest chance for success. If you look at all the top industries mentioned in this book you will note that almost every one of them is engaged in multiplying human efforts. If the products also have a quick payback period, they will be readily accepted.

Price/Earnings Ratio

How much a growth stock costs in comparison to its yearly earnings varies widely. A high tech growth stock that sold for 10 times its earnings during a period of economic expansion would certainly not be reflecting a vote of confidence by the market. On the other hand, a stock that sold for 30 times earnings or more would probably be headed for a bust, although P/Es of 50 and higher are not unheard of.

A stock's P/E must be evaluated in relation to other companies in the industry and the economic outlook. Generally, I look for a company whose P/E is high enough to indicate investor interest but not so high that any little bad news will send prices tumbling. In most favorable markets, a P/E of 15 or 20 as measured against next year's expected earnings is about right.

Debt

Excessive debt has killed more promising companies than any other factor because unexpected events that cause losses make debt service difficult or impossible. Even during good times, payments on high debts reduce income. Therefore, I suggest you select only growth companies with little or no debt.

As it turns out, finding growth companies that are squeaky clean or nearly debt free isn't as difficult as it may appear. Most high tech growth firms raised millions of dollars from their public offerings, which often sustains them well into the period of rapid growth. It isn't unusual to find very young, promising companies with quite a bit of cash and no large obligations.

Return on Equity

You should expect to see at least a 20 percent return on equity for any growing high tech firm. A 20 percent level is necessary to finance most growth internally, thereby eliminating the need to take on debt.

Management

Alas, the bright people that create new technology are often the worst ones to run the company established to produce it. As a result, it is important to see that key officers are experienced in running a business. If they are, and if the founders are also in positions of responsibility to keep them highly motivated, you probably have a winning management team.

It is also worthwhile to see how management is compensated. Officers with large blocks of stock and moderate salaries are committed to making the business succeed if they expect it to make them wealthy. It may be seen as an additional vote of confidence if any of the venture capitalists have retained stock in the firm. (If the company has been public for less than a year or so, request a prospectus for the pertinent information. Otherwise, ask for a 10-K, which is a document somewhat like a prospectus but is published by more mature companies.)

MAKING THE SEARCH

It is one thing to have a good stock-selection checklist, but it is another to put it to use. Most people don't have time to apply selection criteria to more than a few issues. However, if you have access to a personal computer, your problems may be over.

Computers are superb at sorting through large quantities of information and selecting those items that meet specific criteria. In the usual screening process an investor might use, the computer inspects thousands of pieces of information about companies and selects for one characteristic, perhaps annual sales. All the companies that fit the standard are put into a separate group, which is then examined for the next point, let us say the return on equity. The screening process continues until only those issues that fit all the criteria are left. Believe me, once you've seen a computer sort through thousands of companies and find those that meet your standards, you'll become an avid computerized investor.

Institutional investors and mutual fund managers have used computers to screen data for years, which gave them a big edge in the market. However, about a year ago, effective screening tools were developed for IBM and IBM-compatible personal computers that bring powerful computer screening well within the reach of individuals.

Standard & Poor's Stockpak II and Value Line's Value/Screen are both excellent programs. Subscribers are sent a new data disk each month (or several data disks depending on the service requested) that contains fundamental information about thousands of companies. Both programs allow the user to screen stocks using many standard criteria whose values may be selected to suit any needs. The programs are quite easy to use.

Dow Jones's Market Microscope and Savant's Fundamental Investor are similar to the previous programs except there are no monthly datadisks. Instead, users screen information in a national database via a telephone connection.

Using an on-line database instead of datadisks can increase the number of companies available for screening. Often the information is more current in a databank than it might be on a disk. However, telephone and database charges can add up quickly for investors that make heavy use of on-line services. Of course, extensive screens need not be done frequently.

Computers are rapidly becoming essential for modern investors. People who take the trouble to become familiar with the better new investment programs soon find themselves at a great advantage over those who don't. Investment decisions must still be made by people, but computers can make the process easier, more effective, and much more profitable.

Whatever methods you may use in your investment analysis, I'm confident you will find them very rewarding when applied to the selection of high tech growth stocks.

Sources of Computer Investment Products

Company	Product
Dow Jones & Company, Inc. P.O. Box 300 Princeton, N.J. 08540 800/257-5114	Dow Jones Market Microscope
Dow Jones-Irwin, Inc. 1818 Ridge Road Homewood, Ill. 60430 312/798-6000	*Automating Your Financial Portfolio,* by Donald Woodwell
Savant Corporation P.O. Box 440278 Houston, Tex. 77244 713/556-8363	The Fundamental Investor
Standard & Poor's 25 Broadway New York, N.Y. 10004 212/208-8000	Stockpak II
Value Line, Inc. 711 Third Ave. New York, N.Y. 10017 212/687-3965	Value/Screen

CHAPTER 46

Attractive High Tech Mutual Funds: Approached Intelligently, They're Very Profitable

Over the past few years, several mutual funds have been formed to profit from the substantial growth occurring in technology stocks. Many of the better run high tech funds have done quite well and have performance figures running to several hundred percent over just a few years.

Understandably, the top tech funds have attracted quite a bit of attention. Almost every investor who carries the scars of many a Wall Street battle knows someone who put his money in a mutual fund and made a bundle with no effort. Of course, it is satisfying to know that mutual fund investors miss out on most of the fun, but the fact remains that some mutual fund buyers actually do better than "real" investors. As a result, many experienced and capable stock market vets now use high tech mutual funds to round out their portfolios.

HOLDING COMPANIES IN DISGUISE

Mutual funds, for those who are not familiar with their particulars, operate somewhat like holding companies. They accept money from investors and use it to buy shares of promising companies that meet preset criteria that vary according to the objectives of the fund.

The method one uses to buy into a mutual fund depends on how the fund is structured. Most funds deal directly with the public and accept an initial fee, commonly a minimum of $500, to set up the account. Subsequent investments can almost always be made at any time and usually for much less than the opening amount.

Technically when one buys into a mutual fund, one buys stock in the fund. However, investors don't buy individual shares as they would if they purchased the shares of a company. Instead they put in fixed dollar amounts and are assigned however many shares that buys. The value of each individual share, its net asset value (NAV), is how much it would be worth if the fund's portfolio were sold that day and the money distributed to the shareholders. The NAV is also the redemption price per share when one sells them back to the fund.

The net asset value, of course, changes daily as the value of the fund's portfolio changes. As a result, each fund's NAV is reported under Mutual Fund Quotations in *The Wall Street Journal*, making it easy to track performance.

You may notice that the quotations for several mutual funds show an offer price that is somewhat higher than the NAV. Such funds charge a sales commission, called a load, for their shares, which is reflected in the higher prices. Funds listed with an "N.L." next to the net asset value per share are no-load funds that carry no premium.

Load or no-load, almost all mutual funds are open ended; that is, they don't have a fixed number of shares. Investors buy $500 worth of a fund, not 100 shares of a $5 fund. However, there are a few closed-end funds with a fixed number of shares. They are priced according to supply and demand, as are the shares of any other publicly traded company. Closed-end funds frequently trade for more or less than their underlying NAV, sometimes substantially so.

THE ADVANTAGES OF MUTUAL FUNDS

Although the performance of many professional money managers is certainly less than breathtaking, quite a few mutual funds are under the direction of top people with excellent track records, credentials which few individual investors can match. The best managers are well connected in investment circles and can obtain valuable information well ahead of individual investors, or their brokers. Top mutual funds employ competent research staffs and computers to gather and process significant economic and financial data that no individual could possibly process in time to be of value. All in all, the level of management skills available through mutual funds is worth their modest cost, which is usually between 0.5 percent and 1 percent of the fund's assets annually.

The diversified portfolios of mutual funds are also an advantage. I know that Gerald Loeb, the famous Wizard Of Wall Street, said that the way to riches was to very carefully "put all your eggs in one basket and then watch that basket." However, unless you are also a wizard, I suggest you diversify not only for safety but to actually maximize gains. For most people with less than $10,000 to invest, the most cost-effective way to diversify is with a mutual fund.

Speaking of cost effectiveness, funds trade stocks in such large blocks that brokerage fees are well under what an individual would pay even at a discount broker. In fact, brokerage fees are sufficiently low for a fund's trades that they are simply not an important factor in the performance of the fund. That certainly can't be said for an individual's portfolio.

Several of the better technology mutual funds are members of a diverse family of funds offered by the same company. Most such companies allow holders of one fund to switch to any of the other funds, which often includes a money market fund, with a phone call. Although a nominal fee of 1 percent or 2 percent is often charged to discourage frivolous transfers, the service can be extremely useful when the economy or your personal needs change.

Mutual funds are also very easy to rate. Several independent services keep close tabs on the performance of hundreds of funds. In addition, many popular magazines such as *Money* and *Forbes* publish the track records of most funds for investors to evaluate. Because each fund's daily value is published in the financial media, no investor need be in the dark regarding how any fund is doing.

A major advantage to mutual fund investing is that one fund

is far easier to follow than a portfolio of many stocks. By concentrating on one issue, you will probably be able to make a better selection in the first place and do more detailed analysis after the purchase.

Lastly, mutual funds must supply new clients with a prospectus that contains a full disclosure of all negative as well as positive attributes. Included in the prospectus are the contents of the portfolio, a statement of the fund's objectives, the capabilities of management, and other pertinent information.

WHICH FUND IS FOR YOU?

Investors cannot properly select a mutual fund without first examining their own investment objectives. Depending on one's personal needs, the status of the economy, and other factors, an investor can select a fund with any desired risk-to-reward ratio. Most high tech funds fall into one of the three categories.

Maximum Growth Funds

These funds place an emphasis on capital gains and are willing to take greater than average risks to obtain them. Shares of small, emerging technology firms are highly favored due to their potential for appreciation. These are the most speculative of the funds, although some maximum growth portfolios are limited to shares of high flyers that have at least established a short but credible track record.

As may be expected, maximum growth funds are often quite volatile. During periods of market advances that favor tech stocks, such funds can soar. During market declines, however, maximum growth almost always turns to maximum loss, and quickly. These are not funds for the faint of heart.

Long-Term Growth Funds

These funds are a step up the ladder of safety from maximum growth funds. Issues selected for the portfolio are usually from small but well-seasoned companies with established products that enjoy leadership positions in their markets.

Although the companies selected by long-term growth funds are of higher quality than those often found in the portfolios of more aggressive funds, during sharp market declines both will suffer. Sometimes their declines are so great it is difficult to tell them apart. Generally, however, long-term growth funds are less volatile than maximum growth funds. Of course, they are not usually quite as spectacular on the upside either.

Growth and Income Funds

These funds represent a third level of safety and are generally the highest level found with technology funds that are, by nature, primarily capital gains oriented. Growth and income funds are very long-term oriented and balance their portfolios for both dividend income and reasonable capital gains. High tech blue chip companies are frequently found in the portfolios of these funds.

In regard to the choice of no-load funds versus load funds, they have equally good track records overall. However, individual funds often depart from the average. If a fund that carries a sales commission suits your needs and you intend to hold it for a long period, you should not let the added cost deter you from making a purchase. However, make a good search of the no-load funds first.

EVALUATE TRACK RECORDS WITH CARE

The performance records of mutual funds are readily available and should be carefully used when making selections. Unfortunately, however, too many investors simply locate the funds that meet their investment objectives, see which one performed the best the previous year, and send in their money.

In actuality, performance figures must be examined with greater care to be of value. A critical figure for long-range investors is the fund's average performance over several years. The period of time covered should be long enough to include a full economic cycle, with bad periods as well as periods of expansion. Please remember that funds that perform well during bull markets often give up their spectacular gains when the markets decline. As a result, you may be surprised to find that a well-managed fund with a modest average yield will outperform the high fliers over time.

Mutual fund investors who have short-range goals must also do their homework with performance figures. It is important to find the fund's average performance for the type of economic conditions and markets you expect for the duration of your investment. If you are in a market that favors capital-spending stocks, it is well worth while to see how various funds did during similar periods in the past.

Generally, the shorter the range of your investment objectives, the more likely it is that past performance will continue. Portfolios have momentum just like business cycles. Thus, a winning fund is likely to remain a winner for as long as the conditions that made it successful continue.

OTHER SELECTION CRITERIA

Quite often, investors will be urged to critically evaluate the quality of management. I feel you should let the firm's performance record be all the evaluation you need, with one exception. Beware of a fund that has a good record but just changed management. It isn't the same fund anymore, in spite of its having the same name.

The size of the fund can be an important consideration. Because it is difficult to move large amounts of money into the shares of little companies, very aggressive investors should be wary of funds with $100 million or more to move around. On the other hand, the economics of the mutual fund business make it difficult to afford top management if the fund has less than $25 million.

One factor that should not be considered as a basis of selection is the per-share price of mutual funds. Unlike stock in a publicly traded company, where low price can indicate low quality, the shares of mutual funds are sold in dollar amounts as discussed earlier. Mutual funds adjust the number of shares as they see fit but your value will change only as the value of the portfolio changes. Different funds have different ideas regarding how many shares they should use, with the result that equally good funds may have a wide variety of share prices.

MAXIMIZING GAINS

Most well-run mutual funds outperform the market during good times and do worse during bad markets. Because high tech funds tend to be more aggressive, their performance is even more pronounced, in both directions. As a result, it is very worthwhile to switch out of mutual funds during long periods of market declines.

Fortunately, it isn't necessary to pick economic turnaround points in order to profit from the tendency of funds to accentuate market movements. You can be late getting into bull markets and late to get out and still improve your overall gains by an enormous amount. It is only important not to miss major long-term movements. If you are in during most of a recovery and out during most of a recession you will do very well indeed.

At all times, it is a good idea to sell any fund whose performance drops into the bottom half of funds with similar objectives. Recovery may occur, but why wait and perhaps let good market conditions slip by? If you have a no load fund, you have no sales commissions to worry about, although a small redemption fee may be charged. Stick with the winners when the economy is strong and stay out at other times.

CAREFUL INVESTORS MAKE HANDSOME PROFITS

Mutual funds can be very profitable for investors that treat them as conscientiously as they would any other investment. If you do your initial homework, then keep your eye on the performance of your fund and the status of the economy, you will find your investment to be unusually rewarding.

TABLE 1

Selected Technology Mutual Funds

Name	Class*	Phone Switch	Sales* Charge (Percent)	Change			Risk Rating	Phone Number(s)
				1 Year	5 Years	10 Years		
Alliance Technology 140 Broadway New York, N.Y. 10005	LTG	No	8.5	6.54	0.00	0.00	3	800-221-5672

Alliance Technology has a diversified high tech portfolio that primarily, but not exclusively, contains securities of firms with established track records. The fund may invest to a limited extent in warrants, listed call options, restricted securities, and securities of foreign corporations.

Name	Class*	Phone Switch	Sales* Charge (Percent)	1 Year	5 Years	10 Years	Risk Rating	Phone Number(s)
Fidelity Select Technology 82 Devonshire Street Boston, Mass. 02109	LTG	Yes	2	10.24	0.00	0.00	3	617-523-1919 800-225-6190

Select Technology invests in a broad range of high tech companies in many industries. The majority of the portfolio is composed of firms with established track records, although several small, young companies are also present. The fund is a member of a family of eight funds that span diverse sectors.

Name	Class*	Phone Switch	Sales* Charge (Percent)	1 Year	5 Years	10 Years	Risk Rating	Phone Number(s)
IDS Discovery 1000 Roanoke Bldg. Minneapolis, Minn. 55402	Max G	Yes	5	14.30	141.36	0.00	3	612-372-3131 800-437-4332

After the IDS Growth Fund became popular a few years ago, it threatened to grow too unwieldy for investing in small high tech firms. The Discovery Fund was formed and now specializes in small emerging growth technology companies.

TABLE 1 *(continued)*

Name	Class*	Phone Switch	Sales* Charge (Percent)	Change 1 Year	Change 5 Years	Change 10 Years	Risk Rating	Phone Number(s)
IDS Growth Fund 1000 Roanoke Bldg. Minneapolis, Minn. 55402	Max G	Yes	5	6.30	140.61	500.59	4	612-372-3131 800-437-4332

Growth Fund primarily contains all U.S. high tech companies. To be included in the portfolio, each firm must have at least a four-year track record. About 25 percent to 40 percent of the portfolio is composed of companies in the computer and computer-based automation fields.

Name	Class*	Phone Switch	Sales* Charge (Percent)	Change 1 Year	Change 5 Years	Change 10 Years	Risk Rating	Phone Number(s)
IDS New Dimensions 1000 Roanoke Bldg. Minneapolis, Minn. 55402	LTG	Yes	5	15.19	138.67	385.80	3	612-372-3131 800-437-4332

New Dimensions specializes in large technology companies only, which have at least $150 million in annual sales. Approximately 25 percent of the fund is in foreign high tech firms, primarily Japanese.

Name	Class*	Phone Switch	Sales* Charge (Percent)	Change 1 Year	Change 5 Years	Change 10 Years	Risk Rating	Phone Number(s)
Kemper Technology 120 South LaSalle Street Chicago, Ill. 60603	G&I	Yes	8.5	3.66	59.72	326.98	3	312-781-1121 800-621-1048

Unlike most technology funds that concentrate in small, emerging growth companies, Kemper's Technology Fund also includes larger more stable firms, as well as companies that will benefit from using new technology.

Name	Class*	Phone Switch	Sales* Charge (Percent)	Change 1 Year	Change 5 Years	Change 10 Years	Risk Rating	Phone Number(s)
National Aviation & Technology 50 Broad Street New York, N.Y. 10004	LTG	Yes	8.5	20.39	71.29	363.88	3	212-482-8100

National Aviation & Technology is the only mutual fund that specializes in the area of airlines, aerospace/defense, and technology. The portfolio contains a mixture of large, well-established firms and emerging growth companies. Investors may exchange with the National Telecommunications & Technology Fund.

National Telecommunication & Technology Fund† 50 Broad Street New York, N.Y. 10004	LTG	Yes	8.5	—	—	—	—	212-482-8100

Sister fund to the National Aviation & Technology fund for which exchange privileges are offered. Funds are similar except for industry emphasis.

T. Rowe Price New Horizons 100 E. Pratt Street Baltimore, Md. 02110	Max G	Yes	NL	19.76	113.19	418.77	4	301-547-2308 800-638-5660

The fund invests primarily in small, rapidly growing companies that are at an early stage of their development and before they are widely recognized. The fund also invests in companies that offer promise due to new products, new management, or changes in the economy that favor sharply increased earnings. The fund was established 25 years ago and is not exclusively high tech.

Pro Medical Technology 1107 Bethlehem Pike Flourtown, Pa. 19031	LTG	Yes	NL	13.10	83.99	0.00	3	215-836-5000 800-523-0864

The Medical Technology Fund invests almost exclusively in the stocks of medical service and technology companies. The portfolio includes a broad range of issues from small biotech and biomedical firms to well-established pharmaceutical and hospital management companies.

Vanguard Explorer P.O. Box 2600 Valley Forge, Pa. 19482	Max G	No	NL	6.79	135.78	495.03	4	800-523-7025 800-362-0530

Vanguard Explorer invests primarily in small, relatively unseasoned or embryonic companies in technologies and other high-growth fields. The fund's approach to investing assumes commensurately high investment risks in its security selections although those risks are moderated by diversification. Up to 10 percent of the fund's assets may be placed in restricted securities or other assets that have a limited market, but in early 1985, no such securities were held. Holders may exchange with other Vanguard funds, but not by phone.

TABLE 1 ***(concluded)***

Name	Class*	Phone Switch	Sales* Charge (Percent)	Change 1 Year	5 Years	10 Years	Risk Rating	Phone Number(s)
Vanguard Technology† P.O. Box 2600 Valley Forge, Pa. 19482	LTG	Yes	NL	—	—	—	—	800-523-7025 800-362-0530

The Technology Portfolio is a new fund, one of a series of five (others are Energy, Gold and Precious Metals, Health Care, and Service Economy) in its Specialized Portfolios Fund. The Technology Portfolio is heavily oriented toward electronic companies including computer, semiconductor, and telecommunication firms. No more than 5 percent of the portfolio's assets are invested in securities of companies that have less than three years operating history.

*Key to Symbols: Max G = Maximum Growth.
LTG = Long-Term Growth.
G&I = Growth and Income.
NL = No Load/No Sales Charge.

†No risk rating or change figures available.

Numerical Data Courtesy of Schabacker Investment Management, Gaithersburg, Maryland, Publisher of Mutual Fund Newsletters.

CHAPTER 47

Venture Capital Companies Have Undiscovered Potential

As every investor knows, venture capital rewards are often huge. Start-up investors in Apple Computer paid only 9 cents per share for issues that sold for a whopping $22 at the initial public offering. Each share of Convergent Technologies that sold for $13.50 cost venture capitalists only 40 cents. And Tandom Computers turned all of its investor's quarters into $11.50 when the company went public! The list of firms that returned thousands of percent profit to their backers is a very long one indeed.

THERE'S NO NEED TO MISS OUT

Most high tech stock investors have long resented the fact that by the time a company is available in the stock market, the lion's share of the profits have often been made by the venture capitalists that were able to get in on the ground floor. Without really examining the situation, investors usually believe such attractive venture profits are out of reach. They assume such investments are too complex to put together, are impossible to find, are too risky, and are unaffordable in any case. Well... all these assumptions are wrong, but most people just haven't received the news.

Scattered among the lists of stocks in the financial press are a handful of little-noticed venture capital companies (VCCs) that invest both money and management know-how in start-up enterprises. Several of these VCCs have significant investments in publicly traded high tech issues as well. The mix of private to publicly held shares and high tech to non-high tech varies with each VCC.

VCCs ARE UNIQUE

Venture capital companies operate like closed-end mutual funds but without many of the restrictions that limit such funds. VCCs can raise up to $5 million by selling stock to the public but unlike mutual funds, they can invest in the shares of high-tech companies that are not yet publicly traded. Moreover, venture capital companies do not need SEC approval to acquire over 5 percent of an enterprise's stock, as do mutual funds, and may reward its successful managers with up to 20 percent of a portfolio's appreciation.

Venture capital companies have several attractive advantages for investors. They are far less expensive to purchase than venture capital partnerships, which often begin at $250,000. Since they have publicly traded shares, they offer a degree of liquidity that is simply not available with any other venture investment. As a result, you are not locked into your investment for years as is true with limited partnerships, the traditional venture capital investment vehicle.

Another great advantage to a VCC, with its publicly traded shares, is that investors can monitor prices in the stock pages of the newspaper and, like any public company, VCCs publish quarterly and annual reports. Traditional venture capitalists have no such ongoing market evaluation and must often wait for years on pins and needles to learn how much the public will pay for their assets.

Lastly, venture capital companies offer the investor diversified portfolios and professional management, both of which help ensure success in the sometimes volatile high tech area.

BUT IS ANYBODY WATCHING?

Curiously, in spite of their advantages and obvious investment potential, venture capital companies have attracted very little attention. Although the financial press is full of detailed news regarding many of the companies found in VCC portfolios, they rarely mention the VCCs themselves. Thus, we frequently find venture capital companies with quality high tech investments in their portfolios selling below their net asset value per share.

A handful of people do pay serious attention to these issues. One securities dealer I know made money over the years by watching Amdahl. When it made a jump in price he then bought Heizer, which owned a large block of Amdahl's stock. A day or so later Heizer usually moved up a bit to reflect the higher value of its portfolio. The time lag usually allowed him to make small profits. Equivalent situations exist today with many VCCs.

However, the great majority of investors have yet to take much notice of VCCs. Heizer Corporation has apparently tired of waiting for Wall Street to discover them. The company determined that the combined value of its portfolio exceeded the total market value of Heizer's stock. The situation so irritated the Heizer Corporation that it liquidated the company's holdings recently and distributed them to their investors. Ned Heizer, chairman and president of Heizer Corporation, said, "Anytime you do something new or different it takes a while to educate people about it. It may take five years for the market to learn what good investments (VCCs) are."

I don't think it will take anywhere near that long. However, it may be worth recalling Gerald Loeb's advice about finding a good deal on Wall Street. He pointed out that your investment could remain a good deal for longer than you wished to hold it. Therefore, instead of taking positions in this area, I urge you to monitor these stocks for signs of increasing activity. If VCCs begin to move into the limelight, as I expect they will, you should recognize that an excellent investment opportunity is occurring.

VCCs

Company	Symbol	Exchange
Biotech Capital	BITC	OTC

Comments: Biotech Capital is exclusively oriented toward high tech venture capital investments. However, despite the company's name, its portfolio is very broadly based and contains computer and telecommunications holdings as well as biotechnology investments. Definitely a company to watch.

Company	Symbol	Exchange
Capital Southwest	CSWC	OTC

Comments: This company is the largest publicly traded venture capital firm. The company's preference is for firms with an established track record that need both capital and managerial help. Capital Southwest's portfolio is balanced between technology and high growth, nontechnology investments.

Company	Symbol	Exchange
Greater Washington Investors	GWII	OTC

Comments: This VCC's portfolio preference is toward small, developing companies that have a potential for substantial growth. Greater Washington also arranges managerial assistance to companies in its portfolio.

Company	Symbol	Exchange
Narragansett Capital	NARR	OTC

Comments: Narragansett works to provide equity financing to early-stage growth companies. Narragansett is also gaining a good reputation in leveraged buyout financing.

Company	Symbol	Exchange
Nautilus Fund	NTLS	OTC

Comments: Nautilus has a mixed interest in venture capital and growth stocks. The company has many high tech holdings but is not limited to technology investments. Nautilus is often classified as a closed-end mutual fund. Over its nearly six-year life, the fund's increase in net asset value totaled 291 percent, which is between two and three times the S&P 500 and NASDAQ measures.

Company	Symbol	Exchange
Rand Capital	RAND	OTC

Comments: Rand's portfolio is predominantly venture capital oriented and contains several high tech investments.

CHAPTER 48

Technology Index Options Make Price Volatility an Advantage

All seasoned technology investors know that there are many times when high tech stocks move together, either up or down, as if they were a flock of pigeons responding to the same forces. Sometimes, after a prolonged rise in prices, we are almost able to smell a correction in the air and we wish we could use the event to make money. The same is true when, after a long period in the doldrums, we know a high tech rally is overdue and waiting in the wings.

Two years ago a new type of option was created that is ideally suited to experienced investors who feel competent to predict the future activity of whole sectors of stocks. The Pacific Stock Exchange (PSE) launched Technology Index Options, and The American Stock Exchange (AMEX) opened its doors to the Computer Technology Index Options. Neither of the two index-based options have captured much attention, which is unfortunate because, in good hands, they offer many opportunities for profits while risks remain limited.

Pacific Stock Exchange Technology Index (Comparison with Market Indexes)

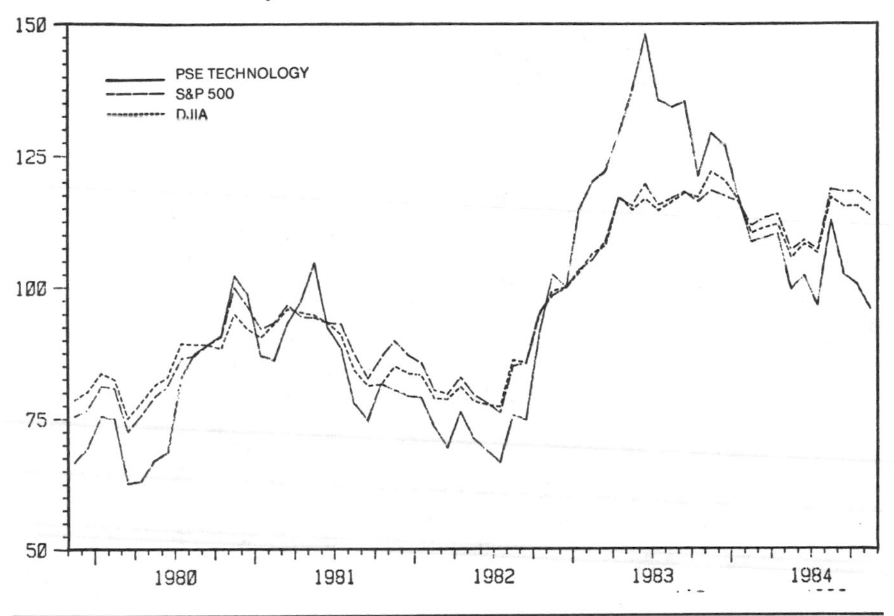

Note: There is no assurance that the same or similar correlations will continue in the future.
Source: The Pacific Stock Exchange, Incorporated.

I'm going to limit my discussion to the PSE Technology Index Options because they are broader in scope than the AMEX version and are perhaps more suitable for most high tech investors. In any event, both the PSE and AMEX options work about the same.

PSE TECHNOLOGY INDEX OPTIONS

In an effort to create an index that closely reflects the price movements of America's diverse high tech industry, a representative group of 100 companies was chosen. The stocks used to calculate the index cover a wide range of technologies and corporate sizes. It is price-weighted so that a price change in one stock will have the same effect on the index as an equivalent price change in any other stock, regardless of capitalization. Although there are some disadvantages to price-weighted indexes, on balance they provide a truer picture of what the *group* of stocks is doing. With a price-weighted index, unusually good or bad news about just one giant company on an otherwise quiet day can't cause a big swing in the index.

In order to make the price movements of the index easier to follow, the index was set at 100 on December 31, 1982. Every day, the index is recalculated and published in *The Wall Street Journal* in the section on index options—the Pacific Stock Exchange is usually listed at the bottom of the column. Many investors who don't trade index options still monitor the index because it is an excellent barometer of the technology sector. Some investors make it a point to plot the index.

Technology Index Options are useful when you feel strongly that tech stocks are due to move significantly in one direction or another. Perhaps you detect signs that a long wave of pessimism has run its course and tech stocks will rebound. In such a case you could purchase a "call," which is an option to buy the value of the index at a particular "strike price" you pick at the time of purchase. Conversely, if your expectations were that tech stocks were about to drop in price, you could purchase a "put," which gives you the right to sell at a particular price. In both cases, strike prices are set five points apart when the index is below 200, and 10 points apart thereafter.

Unlike stocks, options are only valid for a particular amount of time. In the case of index options you buy the right to exercise them for one, two, or three months. If the market moves against you throughout that time, you lose the cost of the option.

On the other hand, if the technology stocks cooperate and the index moves to or beyond your strike price, you have two courses of action available to you. You can call your broker with the instructions that you wish to "exercise the option." The broker will then see that you are paid in cash. The amount you get is determined by taking the difference between the strike price of the option and the index level at the close of trading on the day of the exercise and multiplying it by $100.

Alternately, if your option still has some time on it before it expires, you can ask your broker to sell it. You will probably come out ahead by selling rather than exercising the option because in addition to its intrinsic value, someone will pay you for its remaining "time" value. If the index has been mov-

Pacific Stock Exchange Technology Index (Underlying Stocks)

Advanced Micro Devices, Inc. (AMD)
Altos Computer System (ALTO)
Amdahl Corporation (AMH)
Analog Devices, Inc. (ADI)
Apollo Computer, Inc. (APCI)
Apple Computer, Inc. (AAPL)
Applied Magnetics Corporation (APM)
Applied Materials, Inc. (AMAT)
ASK Computer Systems, Inc. (ASKI)
Avantek, Inc. (AVAK)
AVX Corporation (AVX)
BMC Industries, Inc. (BMC)
Burroughs Corporation (BGH)
Centronics Data Computer Corp. (CEN)
Cetus Corporation (CTUS)
Cipher Data Products, Inc. (CIFR)
Cobe Laboratories, Inc. (COBE)
Commodore International Ltd. (CBU)
Communications Industries, Inc. (COMM)
Communications Satellite Corp. (CQ)
Computervision Corporation (CVN)
Conrac Corporation (CAX)
Control Data Corporation (CDA)
Control Laser Corporation (CLSR)
Convergent Technologies, Inc. (CVGT)
Corvus Systems, Inc. (CRVS)
CPT Corporation (CPTC)
Cray Research, Inc. (CYR)
Cullinet (CUL)
Daisy Systems Corp. (DAZY)
Data I/O Corporation (DAIO)
Data General Corporation (DGN)
Datapoint Corporation (DPT)
Datum, Inc. (DATM)
Digital Equipment Corporation (DEC)
Digital Switch (DIGI)
Electro-Biology, Inc. (EBII)
Electronic Associates, Inc. (EA)
Electronic Memories & Mag. Corp. (EMM)
Enzo Biochem (ENZO)
Evans & Sutherland Computer Corp. (ESCC)
Finnigan Corporation (FNNG)
Floating Point Systems, Inc. (FLP)
Flow General, Inc. (FGN)
Fluke (John) Manufacturing Co., Inc. (FKM)
GCA Corporation (GCA)
Genentech, Inc. (GENE)
General DataComm Industries, Inc. (GDC)
GenRad, Inc. (GEN)
Gould, Inc. (GLD)
Harris Corporation (HRS)
Healthdyne, Inc. (HDYN)
Hewlett-Packard Company (HWP)
Honeywell, Inc. (HON)
Hybritech, Inc. (HYBR)
Intel Corporation (INTC)
Intergraph Corporation (INGR)
Intermedics, Inc. (ITM)
International Business Machines Corp. (IBM)
KLA Instruments Corp. (KLAC)
Kulicke and Soffa Industries (KLIC)
Lexidata Corporation (LEXD)
Lotus Development (LOTS)
Medtronic, Inc. (MDT)
Micom Systems, Inc. (MICS)
Millipore Corporation (MILI)
Modular Computer Systems, Inc. (MSY)
Monolithic Memories, Inc. (MMIC)
Motorola, Inc. (MOT)
National Micronetics,Inc. (NMIC)
National Semiconductor Corp. (NSM)
NBI, Inc. (NBI)
NCR Corporation (NCR)
Network Systems Corporation (NSCO)
Nicolet Instrument Corporation (NIC)
Novo Industri A/S (NVO)
Optical Radiation Corp. (ORCO)
Pansophic Systems (PANS)
Paradyne Corporation (PDN)
Perkin-Elmer Corporation (PKN)
Prime Computer, Inc. (PRM)
Quantum Corporation (QNTM)
Rogers Corporation (ROG)
Scientific-Atlanta, Inc. (SFA)
Seagate Technology (SGAT)
Silicon Systems, Inc. (SLCN)
Spectra-Physics, Inc. (SPY)
Sperry Corporation (SY)
Standard Microsystems Corporation (SMSC)
Tandem Computers Incorporated (TNDM)
Tandon Corporation (TCOR)
Tektronic, Inc. (TEK)
TeleVideo Systems, Inc. (TELV)
Telex Corporation (TC)
Teradyne, Inc. (TER)
Texas Instruments, Inc. (TXN)
Timplex Communications, Inc. (TIX)
Varian Associates, Inc. (VAR)
Verbatim Corporation (VRB)
Wang Laboratories, Inc. (WAN)

Totals:

NYSE	48
AMEX	5
OTC	47

Note: The Pacific Stock Exchange reserves the right to add and/or delete stocks to maintain the integrity of the Index.

Source: The Pacific Stock Exchange, Incorporated.

Pacific Stock Exchange Technology Index (Stock Weightings as of December 13, 1983)

SYMBOL	COMPANY NAME	PERCENT WEIGHT	SYMBOL	COMPANY NAME	PERCENT WEIGHT
MOT	MOTOROLA INC	4.4189	TC	TELEX CORP	0.7506
HON	HONEYWELL INC	4.3906	CVGT	CONVERGENT TECHNOLOGIES INC	0.7466
TXN	TEXAS INSTRS INC	4.3866	DYSN	DYSAN CORP	0.7425
NCR	NCR CORP	4.1929	AAPL	APPLE COMPUTER INC.	0.7264
IBM	INTERNATIONAL BUSINESS MACHS	3.9427	DAIO	DATA I O CORP	0.7103
TEK	TEKTRONIX INC	2.3487	TCOR	TANDON CORP	0.6618
DEC	DIGITAL EQUIP CORP	2.2801	CIFR	CIPHER DATA PRODS INC	0.6538
NVO	NOVO INDUSTRI A S	1.9290	AMH	AMDAHL CORP	0.6215
VAR	VARIAN ASSOC INC	1.7353	BMC	BMC INDS INC MINN	0.6215
CYR	CRAY RESH INC	1.6949	QNTM	QUANTUM CORP	0.6215
BGH	BURROUGHS CORP	1.5981	HYBR	HYBRITECH INC	0.6134
SY	SPERRY CORP	1.5012	PDN	PARADYNE CORP	0.6134
CDA	CONTROL DATA CORP DEL	1.4447	COBE	COBE LABS INC	0.6053
CVN	COMPUTERVISION CORP	1.3842	HDYN	HEALTHDYNE INC	0.6053
MDT	MEDTRONIC INC	1.3680	NSCO	NETWORK SYS CORP	0.6013
INTC	INTEL CORP	1.3640	TELV	TELEVIDEO SYS INC	0.5892
MICS	MICOM SYS INC	1.3640	VRB	VERBATIM CORP	0.5730
HWP	HEWLETT PACKARD CO	1.3237	SLCN	SILICON SYS INC	0.5650
ADI	ANALOG DEVICES INC	1.3196	ITM	INTERMEDICS INC	0.5609
HRS	HARRIS CORP DEL	1.3035	PRM	PRIME COMPUTER INC	0.5609
CBU	COMMODORE INTL LTD	1.2793	ASKI	ASK COMPUTER SYS INC	0.5408
GCA	GCA CORP	1.1864	CAX	CONRAC CORP	0.5327
INGR	INTERGRAPH CORP	1.1864	NSM	NATIONAL SEMICONDUCTOR CORP	0.5327
GEN	GENRAD INC	1.1743	NIC	NICOLET INSTR CORP	0.5246
GENE	GENENTECH INC	1.1461	SFA	SCIENTIFIC ATLANTA INC	0.5125
TER	TERADYNE INC	1.1300	CEN	CENTRONICS DATA COMPUTER CORP	0.5085
FLP	FLOATING POINT SYS INC	1.1299	SGAT	SEAGATE TECHNOLOGY	0.4681
TNDM	TANDEM COMPUTERS INC	1.1259	STR	STORAGE TECHNOLOGY CORP	0.4681
ROG	ROGERS CORP	1.0896	CPTC	C P T CORP	0.4479
DGN	DATA GEN CORP	1.0856	CCTC	COMPUTER & COMMUNICATIONS TECH	0.4358
AMD	ADVANCED MICRO DEVICES INC	1.0654	NMIC	NATIONAL MICRONETICS INC	0.4116
WAN	WANG LABS INC	1.0613	CTUS	CETUS CORP	0.3672
ESCC	EVANS & SUTHERLAND COMPUTER CP	1.0573	EBII	ELECTRO BIOLOGY INC	0.3632
APCI	APOLLO COMPUTER INC	1.0492	CRVS	CORVUS SYS INC	0.3309
AMAT	APPLIED MATLS INC	1.0331	ADAC	ADAC LABS	0.3269
GDC	GENERAL DATACOMM INDS INC	1.0089	ONIX	ONYX IMI INC	0.3269
CQ	COMMUNICATIONS SATELLITE CORP	1.0048	FGN	FLOW GEN INC	0.3228
GLD	GOULD INC	1.0048	ALTO	ALTOS COMPUTER SYSTEMS	0.3148
FKM	FLUKE JOHN MFG INC	0.9726	FNNG	FINNIGAN CORP	0.3148
DPT	DATAPOINT CORP	0.9403	CLSR	CONTROL LASER CORP	0.3027
MMIC	MONOLITHIC MEMORIES INC	0.9120	DATM	DATUM INC	0.2623
MILI	MILLIPORE CORP	0.9040	EMM	ELECTRONIC MEMORIES & MAGNETICS	0.2583
PKN	PERKIN ELMER CORP	0.8959	IDPY	INFORMATION DISPLAYS INC	0.2542
SPY	SPECTRA PHYSICS INC	0.8757	MSY	MODULAR COMPUTER SYS INC	0.2421
APM	APPLIED MAGNETICS CORP	0.8434	LEXD	LEXIDATA CORP	0.2300
GRG	GRANGER ASSOC	0.8394	EA	ELECTRONIC ASSOC INC	0.2260
NBI	NBI INC	0.8394	FSYS	FORTUNE SYS CORP	0.1937
AVX	AVX CORP	0.8313	VCTR	VECTOR GRAPHIC INC	0.0646
SMSC	STANDARD MICROSYSTEMS CORP	0.8273			
TIX	TIMEPLEX INC	0.8111			
COMM	COMMUNICATIONS INDS INC	0.8071			
AVAK	AVANTEK INC	0.7546			

Source: The Pacific Stock Exchange, Incorporated.

ing strongly in your direction and you have lots of time left on the option, you may find that the time value will be very high. But don't dither around trying to make up your mind. The time value gets lower as the time remaining on the option gets shorter.

Index options can be useful to hedge portfolios, but only to a limited extent. If you have a large portfolio that is well represented with technology stocks, you could partly protect them from a sudden drop in price by using index options. However, it is important to note that no index will reflect your personal portfolio so closely that it can be totally effective as a hedge. In fact, the better you are as an investor, the less useful they may be. For example, tech stocks in general (and, of course, the Tech Stock Index) could be moving sharply down while your well-chosen portfolio was holding its own.

Index options can be purchased on margin, although I recommend against doing so. Index options are already highly leveraged vehicles. To leverage them further boosts the risks above what I feel their potential value merits.

All investors who are interested in trading options should read the following two booklets: *Understanding The Risks and Uses of Listed Options* and *Listed Options on Stock Indices*. They are free from the Options Clearing Corporation, 200 South Wacker Drive, 27th Floor, Chicago, Ill. 60606, or you may obtain them from your broker. Both booklets are excellent.

There are many sophisticated techniques available for options traders that readers may enjoy investigating. Some of the most elaborate trading methods involve the use of computers. However, from my observations over many years, most people who use options to great advantage simply buy puts and calls on those rare occasions when they are very sure of themselves. At other times, they buy and sell the stocks themselves. As one overly modest investor explained to me recently, "With options you must be right about your choice and also right about the timing. I can never get them both right consistently so I'll just have to get rich slower than all those smart fellows." With that, he dipped his head and slowly sat back—in his chauffeur-driven Mercedes.

Part III

Following Up on Opportunities

Section 12

A Directory of Major Public High Tech Companies

CHAPTER 49

Using This Directory

Readers should find this reference section to be a valuable source of contact information for the high tech companies reviewed in this book. In addition, the directory will be quite useful to serious investors who wish to obtain additional information about public technology companies they see mentioned in the news.

The directory will also help investors to follow stock prices using electronic quotation machines and personal computers. Each company's symbol and the exchange where its stock is traded is listed, along with the firm's address. In the directory, NASDAQ refers to the U.S. Over The Counter (OTC) "exchange." NYSE and ASE refer to the New York Stock Exchange and the American Stock Exchange, respectively. Wherever appropriate, each entry includes the company's telephone number.

All the companies in this directory have a significant number of high tech products or are in the process of developing such products. The latter group includes many firms that are quietly changing from conventional businesses into technology-based companies.

Several private companies are listed whose initial public offerings (IPOs) are pending. Note that the stock symbols for these companies have only tentative status and may change.

The foreign section of the directory lists many companies whose stocks are traded in the United States either directly or by means of American depository receipts (ADRs). Although many foreign firms have U.S. sales offices, I list their home office addresses when the latter are the best places to obtain annual reports. Occasionally, I show the telephone number for the U.S. sales office next to the foreign address if both may be used to obtain financial information.

I've made a special effort to include the top Canadian high tech companies in the foreign directory because many of them are quite attractive, are easily traded, and yet are almost unknown in the United States. A few Canadian companies are traded on U.S. exchanges. Please note, however, that the designation "OTC" with a Canadian firm refers to the Canadian Over The Counter market.

To obtain a company's annual report, quarterly report, or other financial information, you may write or call the corporate treasurer or shareholder relations department. Your request need not be elaborate but should include the information that you are considering the purchase of shares in the company.

This directory has over 1,250 entries. Although every attempt has been made to be accurate, some errors undoubtedly exist. Companies change locations, are absorbed by other firms, and occasionally go out of business. In addition, stock symbols change when firms change names or exchanges. If you are unable to contact any company *in this directory* using the listed information, please feel free to drop me a note and I will try to track it down for you.

MAJOR U.S. PUBLIC HIGH TECH COMPANIES

202 DATA SYSTEMS, INCORPORATED
TOOT NASDAQ
1275 DRUMMER LANE
WAYNE PA 19087
215/964-1170

21ST CENTURY ROBOTICS, INC.
TROB NASDAQ
1 PARK PLACE SUITE 204
ATLANTA GA 30318
404/351-6767

3COM CORPORATION
COMS NASDAQ
1365 SHOREBIRD WAY
MOUNTAIN VIEW CA 94043
415/961-9602

ABM COMPUTER SYSTEMS
ABMC NASDAQ
23362 PERALTA DRIVE
LAGUNA HILLS CA 92653
714/859-6531

ACCURAY CORPORATION
ACRA NASDAQ
P.O.BOX 02248
COLUMBUS OH 43202
614/261-2000

ACME-CLEVELAND CORPORATION
AMT NYSE
P.O. BOX 5617
CLEVELAND OH 44101
216/431-4500

ACRIAN, INCORPORATED
ACRN NASDAQ
10060 BUBB ROAD
CUPERTINO CA 95014
408/996-8522

ACRO ENERGY CORPORATION
ACRO NASDAQ
2006 S. BAKER AVENUE
ONTARIO CA 91761
714/947-4100

ACTIVISION, INCORPORATED
AVSN NASDAQ
2350 BAYSHORE FRONTAGE ROAD
MOUNTAIN VIEW CA 94043
415/960-0410

ADAC LABORATORIES
ADAC NASDAQ
225 SAN GERONIMO WAY
SUNNYVALE CA 94086
408/736-1101

ADAGE, INCORPORATED
ADGE NASDAQ
ONE FORTUNE DRIVE
BILLERICA MA 01821
617/667-7070

ADDISON-WESLEY PUBISH. CO.,INC.
ADSNB NASDAO
READING MA 01867
617/944-3700

ADVANCE CIRCUITS, INCORPORATED
ADVC NASDAQ
15102 MINNETONKA INDUSTRIAL RD
MINNETONKA MN 55343
612/935-3311

ADVANCED COMPUTER TECH. CORP.
ACTP NASDAQ
16 EAST 32ND STREET
NEW YORK NY 10016
212/696-3600

ADVANCED GENETIC SCIENCES, INC.
AGSI NASDAQ
666 STEAMBOAT ROAD
GREENWICH CT 06830
203/622-6552

ADVANCED MEDICAL IMAGING CORP.
AMIC NASDAQ
111 GREAT NECK ROAD
GREAT NECK NY 11021
516/466-4970

ADVANCED MICRO DEVICES, INC.
AMD NASDAQ
901 THOMPSON PLACE
SUNNYVALE CA 94086
408/732-2400

ADVANCED NMR SYSTEMS, INC.
ANMR NASDAQ
30 SONAR DRIVE
WOBURN MA 01801
617/938-6046

ADVANCED SYSTEMS, INCORPORATED
ASY NYSE
155 EAST ALGONQUIN ROAD
ARLINGTON HTS IL 60005
312/981-1500

AEGIS MEDICAL SYSTEMS, INC.
AGIS NASDAQ
137 GAITHER DRIVE
MT LAUREL NJ 08054
609/234-4735

AEL INDUSTRIES
AELNA NASDAQ
P.O. BOX 552
LANSDALE PA 19446
215/822-2929

AFP IMAGING CORPORATION
AFPC NASDAQ
50 EXECUTIVE BLVD.
ELMSFORD NY 10523
914/592-6100

AGRIGENETICS CORPORATION
AGTC NASDAQ
3375 MITCHELL LANE
BOULDER CO 80301
303/443-5900

AGS COMPUTERS, INCORPORATED
AGSC NASDAQ
1135 SPRUCE DRIVE
MOUNTAINSIDE NJ 07092
201/654-4321

AIM TELEPHONE, INCORPORATED
AIMT NASDAQ
1275 BLOOMFIELD AVENUE
FAIRFIELD NJ 07006
201/227-9169

ALAN-TOL INDUSTRIES, INC.
ALOT NASDAQ
600 E. GREENWICH AVE.
WEST WARWICK RI 02893
401/828-4000

ALDEN ELECTRONIC CO., INC.
ADN BOSTON
ALDEN RESEARCH CENTER
WESTBORO MA 01581
617/366-8851

ALGOREX CORPORATION
ALGO NASDAQ
6901 JERICHO TURNPIKE
SYOSSET NY 11791
516/921-7600

ALLIED CAPITAL CORPORATION
ALLC NASDAQ
1625 I STREET, N.W.
WASHINGTON DC 20006
202/331 1112

ALLNET COMM. SERVICES, INC.
ALNT NASDAQ
100 SOUTH WACKER DRIVE
CHICAGO IL 60606
312/443-1444

ALPHA INDUSTRIES, INCORPORATED
AHA ASE
20 SYLVAN ROAD
WOBURN MA 01801
617/935-5150

ALPHA MICROSYSTEMS
ALMI NASDAQ
17881 SKY PARK NORTH
IRVINE CA 92714
714/957-1404

ALTOS COMPUTER SYSTEMS
ALTO NASDAQ
2360 BERING DRIVE
SAN JOSE CA 95131
408/946-6700

ALTRON, INCORPORATED
ALRN NASDAQ
ONE JEWEL DRIVE
WILMINGTON MA 01887
617/658-5800

AMCA INTERNATIONAL, LIMITED
AIL TORONTO
DARTMOUTH NATIONAL BANK BLDG
HANOVER NH 03755
603/643-5454

AMDAHL CORPORATION
AMH ASE
1250 EAST ARQUES AVENUE
SUNNYVALE CA 94086
408/746-6000

AMERICAN CELLULAR NETWORK CORP.
AMLL NASDAQ
1 SOUTH NEW YORK AVE STE 802
ATLANTIC CITY NJ 08401
609/348-1400

AMERICAN CYTOGENETICS, INC.
ACYT NASDAQ
6440 COLDWATER CANYON AVE
NORTH HOLLYWOOD CA 91606
213/985-1730

AMERICAN DISTRICT TELEGRAPH
ADT NYSE
ONE WORLD TRADE CENTER
NEW YORK NY 10048
212/558-1100

AMERICAN ELECTROMEDICS CORP.
AECO NASDAQ
98 CUTTER MILL ROAD
GREAT NECK NY 11021
516/466-5100

AMERICAN EDUCATIONAL COMPUTER
AEDC NASDAQ
2450 EMBARCADERO WAY
PALO ALTO CA 94303
415/494-2021

AMERICAN FUEL TECHNOLOGIES
AFTI NASDAQ
P.O. BOX 369
FEDERALSBURG MD 21632
301/754-5061

AMERICAN FIBER OPTICS CORP.
FIBR NASDAQ
1196 E. WILLOW STREET
SIGNAL HILL CA 90806
213/420-1001

AMERICAN GASOHOL REFINERS
AGRA NASDAQ
1512 N. BROADWAY
WICHITA KS 67214
316/263-1567

AMERICAN MEDICAL ALERT CORP.
AMAC NASDAQ
3265 LAWSON BLVD.
OCEANSIDE NY 11572
516/536-5850

AMERICAN MAGNETICS CORPORATION
AMMG NASDAQ
13535 VENTURA BLVD, STE 205
SHERMAN OAKS CA 91423
213/783-8900

AMERICAN NUCLEONICS COPORATION
AMNU NASDAQ
696 HAMPSHIRE ROAD
WESTLAKE VILLAGE CA 91359
805/496-2405

AMERICAN PHYSICIANS SERVICE INC
AMPH NASDAQ
2505 TURTLE CREEK BLVD
DALLAS TX 75219
214/559-4800

AMERICAN SOLAR KING CORPORATION
AMSK NASDAQ
P.O. DRAWER 7399
WACO TX 76714
817/776-3860

AMERICAN SOFTWARE, INC.
AMSWA NASDAQ
443 E. PACES FERRY RD, N.E.
ATLANTA GA 30305
404/261-4381

AMETEK, INCORPORATED
AME NYSE
410 PARK AVENUE
NEW YORK NY 10022
212/935-8640

AMGEN
AMGN NASDAQ
1900 OAK TERRACE LN
THOUSAND OAKS CA 91320
805/499-3617

AMHERST ASSOCIATES, INC.
AASI NASDAQ
20 N. CLARK STREET
CHICAGO IL 60602
312/332-0500

AMI SYSTEMS, INCORPORATED
AMIS NASDAQ
2300 RUTLAND DRIVE
AUSTIN TX 78758
512/836-1700

AMISTAR CORPORATION
AMTA NASDAQ
2675 SKYPARK DRIVE
TORRANCE CA 90505
213/539-7200

AMP, INCORPORATED
AMP NYSE
EISENHOWER BLVD
HARRISBURG PA 17105
717/564-0100

AMPLICA, INCORPORATED
AMPI NASDAQ
950 LAWRENCE DRIVE
NEWBURY PARK CA 91320
805/498-9671

AN-CON GENETICS, INCORPORATED
ANCN NASDAQ
1 HUNTINGTON QUADRANGLE
MELVILLE NY 11747
516/694-8470

ANACOMP, INCORPORATED
AAC NYSE
11550 N. MIRIDIAN, STE 600
CARMEL IN 46032
317/844-9666

ANALOG DEVICES, INCORPORATED
ADI NYSE
RT 1, INDUSTRIAL PARK
NORWOOD MA 02062
617/329-4700

ANALOGIC CORPORATION
ALOG NASDAQ
AUDUBON ROAD
WAKEFIELD MA 01880
617/246-0300

ANALYSTS INT'L CORPORATION
ANLY NASDAQ
7615 METRO BLVD
MINNEAPOLIS MN 55435
612/835-2330

ANAREN MICROWAVE, INCORPORATED
ANEN NASDAQ
6635 KIRKVILLE ROAD
EAST SYRACUSE NY 13057
315/432-8994

ANDERSON 2000, INCORPORATED
ANDN NASDAQ
P.O. BOX 20769
ATLANTA GA 30320
404/997-2000

ANDERSON GROUP, INCORPORATED
ANDR NASDAQ
1280 BLUE HILLS AVENUE
BLOOMFIELD CT 06002
203/242-0761

ANDERSON JACOBSON, INCORPORATED
AJ ASE
521 CHARCOT AVENUE
SAN JOSE CA 95131
408/263-8520

ANDOVER CONTROLS CORPORATION
ANDO NASDAQ
YORK AND HAVERHILL STREETS
ANDOVER MA 01810
617/470-0555

ANDREW CORPORATION
ANDW NASDAQ
10500 W. 153RD STREET
ORLAND PARK IL 60462
312/349-3300

ANDROBOT, INCORPORATED
ABOT NASDAQ
101 EAST DAGGETT DRIVE
SAN JOSE CA 95131
408/262-8676

ANDROS ANALYZERS, INCORPORATED
ANDY NASDAQ
2332 FOURTH STREET
BERKELEY CA 94710
415/849-1377

ANTHEM ELECTRONICS
ATM NYSE
174 COMPONENT DRIVE
SAN JOSE CA 95131
408/946-8000

APOGEE ROBOTICS, INC.
APGE NASDAQ
2290 EAST PROSPECT RD.
FORT COLLINS CO 80525
303/221-1122

APOLLO COMPUTER, INCORPORATED
APCI NASDAQ
15 ELIZABETH DRIVE
CHELMSFORD MA 01824
617/256-6600

APPLE COMPUTER, INCORPORATED
AAPL NASDAQ
10260 BANDLEY DRIVE
CUPERTINO CA 95014
408/996-1010

APPLIED BIOSYSTEMS, INC.
ABIO NASDAQ
850 LINCOLN CENTRE DRIVE
FOSTER CITY CA 94404
415/570-6667

APPLIED CAPITAL CORPORATION
ALLC NASDAQ
1625 I STREET NW SUITE 603
WASHINGTON DC 20006
202/331-1112

APPLIED CIRCUIT TECH. INC.
ACRT NASDAQ
2931 LAJOLLA STREET
ANAHEIM CA 92806
714/632-9230

APPLIED COMMUNICATIONS, INC.
ACIS NASDAQ
206 SOUTH 108TH AVENUE
OMAHA NE 68154
402/330-3732

APPLIED DATA RESEARCH, INC.
ADR.B ASE
RT 206 AND ORCHARD RD CN-8
PRINCETON NJ 08540
201/874-9100

APPLIED DNA SYSTEMS, INC.
ADNA NASDAQ
4415 FIFTH AVENUE
PITTSBURGH PA 15213
412/683-1151

APPLIED MAGNETICS CORPORATION
APM NYSE
75 ROBIN HILL ROAD
GOLETA CA 93117
805/964-4881

APPLIED MATERIALS, INCORPORATED
AMAT NASDAQ
3050 BOWERS AVENUE
SANTA CLARA CA 95051
408/727-5555

APPLIED MEDICAL DEVICES, INC.
AMDI NASDAQ
7346 SOUTH ALTON WAY
ENGLEWOOD CO 80112
303/773-6513

APPLIED SOLAR ENERGY CORP.
SOLR NASDAQ
P.O. BOX 1212
CITY OF INDUSTRY CA 91749
213/968-6581

ARCHIVE CORPORATION
ACHV NASDAQ
3540 CADILLAC AVENUE
COSTA MESA CA 92626
714/641-0279

ARGOSYSTEMS, INCORPORATED
ARGI NASDAQ
884 HERMOSA COURT
SUNNYVALE CA 94086
408/737-2000

ARRAYS, INCORPORATED
ARAY NASDAQ
11223 SOUTH HINDRY AVENUE
LOS ANGELES CA 90045
213/410-3977

ARROW ELECTRONICS, INCORPORATED
ARW NYSE
600 STEAMBOAT ROAD
GREENWICH CT 06830
203/622-9030

ASHTON-TATE
TATE NASDAQ
10150 WEST JEFFERSON BLVD
CULVER CITY CA 90230
213/204-5570

ASK COMPUTER SYSTEMS, INC.
ASKI NASDAQ
730 BISTEL DRIVE
LOS ALTOS CA 94022
415/969-4442

ASSOCIATED COMMUNICATIONS CORP.
SCCM NASDAQ
200 GATEWAY TOWERS
PITTSBURGH PA 15222
412/281-1907

AST RESEARCH, INC.
ASTA NASDAQ
2121 ALTON AVENUE
IRVINE CA 92714
714/863-1333

ASTRADYNE COMPUTER IND., INC.
ACII NASDAQ
1122 FRANKLIN AVENUE
GARDEN CITY NY 11530
516/742-9500

ASTRO-MED, INCORPORATED
ASTI NASDAQ
600 EAST GREENWICH AVE.
WEST WARWICK RI 02893
401/828-4000

ASTROCOM CORPORATION
ACOM NASDAQ
120 WEST PLATO BLVD
ST PAUL MN 55107
612/227-8651

ASTRONICS CORPORATION
ATRO NASDAQ
80 S. DAVIS STREET
ORCHARD PARK NY 14127
716/662-6640

ASTROSYSTEMS, INCORPORATED
ASTR NASDAQ
6 NEVADA DRIVE
LAKE SUCCESS NY 11042
516/328-1600

ATLANTIC RESEARCH CORPORATION
ATRC NASDAQ
5390 CHEROKEE AVENUE
ALEXANDRIA VA 22314
703/642-4000

AUGAT, INCORPORATED
AUG NYSE
P O BOX 448
MANSFIELD MA 02048
617/543-4300

AUSTIN, INCORPORATED
ATRN NASDAQ
NORTH INTERSTATE 35
AUSTIN TX 78761
512/251-2341

AUSTRON, INCORPORATED
ATRN NASDAQ
P.O. BOX 14766
AUSTIN TX 78761
512/251-2341

AUTO-TROL TECHNOLOGY CORP.
ATTC NASDAQ
12500 N. WASHINGTON STREET
DENVER CO 80233
303/452-4919

AUTO/RECOGNITION SYSTEMS, INC.
ARES NASDAQ
1132 MARK AVENUE
CARPINTERIA CA 93013
805/684-7715

AUTODYNAMICS, INCORPORATED
AUDYB NASDAQ
300 HALLS MILL ROAD
FREEHOLD NJ 07728
201/462-9400

AUTOMATED SYSTEMS, INC.
ASII NASDAQ
1505 COMMERCE AVENUE
BROOKFIELD WI 53005
414/784-6400

AUTOMATED MEDICAL LABS, INC.
AUML NASDAQ
8405 NORTHWEST 53 ST
MIAMI FL 33166
305/592-1346

AUTOMATED PROFESSIONAL SYSTEMS
ABSI NASDAQ
270 MADISON AVENUE
NEW YORK NY 10016
212/725-2442

AUTOMATIC DATA PROCESSING, INC.
AUD NYSE
405 ROUTE 3
CLIFTON NJ 07015
201/365-7300

AUTOMATIX, INCORPORATED
AITX NASDAQ
1000 TECH PARK DRIVE
BILLERICA MA 01821
617/668-7900

AUTOTROL CORPORATION
AUTR NASDAQ
1701 WEST CIVIC DRIVE
MILWAUKEE WI 53209
414/228-9100

AUXTON COMPUTER (AUXCO), INC.
AUXT NASDAQ
201 EAST PINE STREET
ORLANDO FL 32801
305/425-3300

AVANT-GARDE COMPUTING, INC.
AVGA NASDAQ
2091 SPRINGDALE ROAD
CHERRY HILL NJ 08003
609/424-9620

AVANTEK, INCORPORATED
AVAK NASDAQ
3175 BOWERS AVENUE
SANTA CLARA CA 95051
408/727-0700

AVNET, INCORPORATED
AVT NYSE
767 FIFTH AVENUE
NEW YORK NY 10153
212/644-1050

AVX CORPORATION
AVX NYSE
60 CUTTER MILL ROAD
GREAT NECK NY 11021
516/829-8500

AW COMPUTER SYSTEMS, INC.
AWCSA NASDAQ
9000A COMMERCE PKWY
MT LAUREL NJ 08054
609/234-3939

AYDIN COMPANY
AYD NYSE
401 COMMERCE DRIVE
FT. WASHINGTON PA 19034
215/643-7500

BAIRD CORPORATION
BATM NASDAQ
125 MIDDLESEX TURNPIKE
BEDFORD MA 01730
617/276-6000

BALANCE COMPUTER CORPORATION
BCCO NASDAQ
EXECUTIVE PLAZA 1, STE 400
HUNT VALLEY MD 21031
301/667-1400

BALLY MANUFACTURING CORPORATION
BLY NYSE
2640 W BELMONT AVENUE
CHICAGO IL 60618
312/267-6060

BARNES ENGINEERING COMPANY
BIR ASE
P.O. BOX 53
STAMFORD CT 06904
203/348-5381

BARRINGER RESOURCES, INC.
BARRC NASDAQ
1626 COLE BLVD
GOLDEN CO 80701
303/232-8811

BASE TEN SYSTEMS, INCORPORATED
BASEB NASDAQ
1 ELECTRONICS DRIVE
TRENTON NJ 08619
609/586-7010

BAXTER TRAVENOL LABORATORIES
BAX NYSE
1 BAXTER PARKWAY
DEERFIELD IL 60015
312/948-2000

BEDFORD COMPUTER CORPORATION
BEDF NASDAQ
TIRRELL HILL ROAD
BEDFORD NH 03102
603/668-3400

BEEHIVE INTERNATIONAL
BHI ASE
P.O. BOX 25668
SALT LAKE CITY UT 84125
801/355-6000

BEL FUSE, INCORPORATED
BELF NASDAQ
198 VAN VORST ST.
JERSEY CITY NJ 07302
201/432-0463

BELL & HOWELL COMPANY
BHW NYSE
7100 MCCORMICK ROAD
CHICAGO IL 60645
312/673-3300

BELL INDUSTRIES, INCORPORATED
BI NYSE
11812 SAN VINCENTE BLVD
LOS ANGELES CA 90049
213/826-6778

BESICORP GROUP, INCORPORATED
BESI NASDAQ
BOX 191
ELLENVILLE NY 12428
914/647-6700

BETZ LABORATORIES, INCORPORATED
BETZ NASDAQ
4636 SOMERTON ROAD
TREVOSE PA 19047
215/355-3300

BFI COMMUNICATIONS SYSTEMS, INC
BFXC NASDAQ
109 N. GENESEE STREET
UTICA NY 13502
315/735-9526

BGS SYSTEMS, INCORPORATED
BGSS NASDAQ
29 SAWYER ROAD
WALTHAM MA 02254
617/891-0000

BILLINGS CORPORATION
BIEN NASDAQ
18600 E. 37TH TERRACE
INDEPENDENCE MO 64057
816/373-0000

BIO RECOVERY TECHNOLOGY
8580 KATY FREEWAY, STE 200
HOUSTON TX 77024
713/932-7497

BIO-ANALYTIC LABORATORIES, INC.
BIAL NASDAQ
PO BOX 333 TOM CITY
TAMPA FL 33490
305/287-3340

BIO-LOGIC SYSTEMS CORP.
BLSC NASDAQ
425 HUEHL ROAD
NORTHBROOK IL 60062
312/564-8202

BIO-MEDICUS INCORPORATED
BMDS NASDAQ
15306 INDUSTRIAL ROAD
MINNETONKA MN 55343
612/938-7600

BIO-RAD LABORATORIES, INC.
BIO.B ASE
2200 WRIGHT AVENUE
RICHMOND CA 94804
415/232-4130

BIO-RESPONSE, INCORPORATED
BIOR NASDAQ
550 RIDGEFIELD ROAD
WILTON CT 06897
203/762-0331

BIO-TECHNOLOGY GENERAL CORP.
BTGC NASDAQ
280 PARK AVENUE
NEW YORK NY 10017
212/986-3010

BIOCHEM INTERNATIONAL, INC.
BIOC NASDAQ
P.O. BOX 13157
MILWAUKEE WI 53213
414/542-3100

BIOMET, INCORPORATED
BMET NASDAQ
P.O. BOX 587
WARSAW IN 46580
219/267-6639

BIOSEARCH MEDICAL PRODUCTS, INC.
BMPI NASDAQ
P.O. BOX 1700
SOMERVILLE NJ 08876
201/722-5000

BIOSENSOR CORPORATION
BSNR NASDAQ
2700 FREEWAY BOULEVARD
MINNEAPOLIS MN 55430
612/560-8642

BIOSPHERICS, INCORPORATED
BINC NASDAQ
4928 WYACONDA ROAD
ROCKVILLE MD 20852
301/770-7700

BIOSTIM, INCORPORATED
BIOZ NASDAQ
P.O. BOX 3138
PRINCETON NJ 08540
800/257-5184

BIOTECH CAPITAL CORPORATION
BITC NASDAQ
11 HANOVER SQUARE
NEW YORK NY 10005
212/758-7722

BIOTECH RESEARCH LABS., INC.
BTRL NASDAQ
1600 E. GUDE ROAD
ROCKVILLE MD 20850
301/251-0800

BIOTECHNICA INTERNATIONAL, INC.
BIOT NASDAQ
85 BOULTON STREET
CAMBRIDGE MA 02140
617/864-0040

BIOTECHNOLOGY DEVELOPMENT CORP.
BIOD NASDAQ
400-1 TOTTEN POND ROAD
WALTHAM MA 02154
617/890-0018

BIRDFINDER CORPORATION
BFTV NASDAQ
1999 LINCOLN DR. STE 201
SARASOTA FL 33577
813/955-9280

BIRTCHER CORPORATION (THE)
BIRT NASDAQ
4501 N. ARDEN DRIVE
EL MONTE CA 91731
213/575-8144

BISHOP GRAPHICS, INCORPORATED
BGPH HASDAQ
5388 STERLING CENTER DRIVE
WESTLAKE VILLAGE CA 91359
213/991-2600

BLISS (A.T.) AND CO., INC.
ATBL NASDAQ
1300 N. ANDREWS EXTENSION
POMPANO BEACH FL 33060
305/785-4000

BOEING COMPANY (THE)
BA NYSE
7755 EAST MARGINAL WAY SOUTH
SEATTLE WA 98108
206/655-2121

BOLT BERANEK AND NEWMAN, INC.
BBN AMEX
10 MOULTON STREET
CAMBRIDGE MA 02238
617/491-1850

BOMED MEDICAL MANUF. LIMITED
BOMD NASDAQ
130 MCCORMICK AVENUE, STE 109
COSTA MESA CA 92626
714/754-0305

BOOLE & BABBAGE, INC.
BOOL NASDAQ
510 OAKMEAD PARKWAY
SUNNYVALE CA 94086
408/735-9550

BOSTON DIGITAL CORPORATION
BOST NASDAQ
GRANITE PARK
MILFORD MA 01757
617/473-4561

BOWMAR INSTRUMENT CORPORATION
BOM ASE
850 LAWRENCE DRIVE
NEWBURY PARK CA 91320
805/498-2161

BPI SYSTEMS, INCORPORATED
BPII NASDAQ
3423 GUADALUPE
AUSTIN TX 78705
512/454-2801

BR COMMUNICATIONS
BRHF NASDAQ
1249 INNSBRUCK DRIVE
SUNNYVALE CA 94086
408/734-1600

BRADFORD NATIONAL CORPORATION
BDR ASE
67 BROAD STREET
NEW YORK NY 10004
212/530-2400

BRITTON LEE, INCORPORATED
BRLE NASDAQ
14600 WINCHESTER BLVD
LOS GATOS CA 95030
408/378-7000

BROWNING-FERRIS IND., INC.
BFI NYSE
P.O. BOX 3157
HOUSTON TX 77001
713/870 8100

BSD MEDICAL CORPORATION
BSDM NASDAQ
420 CHIPETA WAY
SALT LAKE CITY UT 84108
801/582-5550

BURNDY CORPORATION
BDC NYSE
RICHARDS AVENUE
NORWALK CT 06856
203/838-4444

BURR-BROWN CORPORATION
BBRC NASDAQ
6730 S. TUCSON BLVD.
TUCSON AZ 85706
602/746-1111

BURROUGHS CORPORATION
BGH NYSE
BURROUGH'S PLACE
DETRIOT MI 48232
313/972-7000

BURST AGRITECH, INCORPORATED
BRZT NASDAQ
6811 WEST 63RD ST, STE 304
OVERLAND PARK KS 66202
613/262-2444

BUSINESSLAND
BUSL NASDAQ
3600 STEVENS CREEK BLVD
SAN JOSE CA 96117
408/554-9300

BYTE INDUSTRIES, INCORPORATED
BYTE NASDAQ
21130 CABOT BOULEVARD
HAYWARD CA 94545
415/783-8272

C-COR ELECTRONICS, INCORPORATED
CCBL NASDAQ
60 DECIBEL ROAD
STATE COLLEGE PA 16801
814/238-2461

C3, INCORPORATED
CCCI NASDAQ
11425 ISSAC NEWTON SQ. S.
RESTON VA 22090
703/471-6000

CADEC SYSTEMS, INCORPORATED
KDCK NASDAQ
410 GREAT ROAD
LITTLETON MA 01460
617/486-8955

CALIFORNIA BIOTECHNOLOGY, INC.
CBIO NASDAQ
2450 BAYSHORE FRONTAGE ROAD
MOUNTAIN VIEW CA 94043
415/966-1550

CALIFORNIA MICROWAVE, INC.
CMIC NASDAQ
990 ALMANOR AVENUE
SUNNYVALE CA 94086
408/732-4000

CAMBEX CORPORATION
CBEX NASDAQ
360 SECOND AVENUE
WALTHAM MA 02154
617/890-6000

CAMBRIDGE BIOSCIENCE CORP.
CBCX NASDAQ
35 SOUTH STREET
HOPHINTON MA 01748
617/435-9071

CAMBRIDGE MEDICAL TECHNOLOGY
CMTC NASDAQ
575 MIDDLESEX TURNPIKE
BILLERICA MA 01865
617/935-4050

CAPITAL SOUTHWEST CORPORATION
CSWC NASDAQ
12900 PRESTON RD AT LBJ
DALLAS TX 75230
214/233-8242

CARDIO-PACE MEDICAL, INC.
CPMI NASDAQ
2833 N. FAIRVIEW AVE.
ST. PAUL MN 55113
612/483-6787

CELL PRODUCTS, INCORPORATED
CELL NASDAQ
5 GEORGES ROAD
NEW BRUNSWICK NJ 08901
201/828-6100

CELLULAR PRODUCTS, INCORPORATED
CELP NASDAQ
688 MAIN STREET
BUFFALO NY 14202
716/842-6270

CELLULAR PRODUCTS, INCORPORATED
CELP NASDAQ
688 MAIN STREET
BUFFALO NY 14202
716/842-6270

CENTOCOR, INCORPORATED
CNTO NASDAQ
244 GREAT VALLEY PARKWAY
MALVERN PA 19355
215/296-4488

CENTRONICS DATA COMPUTER CORP.
CEN NYSE
ONE WALL STREET
HUDSON NH 03051
603/883-0111

CENTURI, INCORPORATED
CENT NASDAQ
245 W. 74TH PLACE
HIALEAH FL 33014
305/558-5200

CERADYNE, INCORPORATED
CRDN NASDAQ
16781-A MILLIKEN AVENUE
IRVINE CA 92714
714/549-0421

CERBERONICS, INCORPORATED
CRBRA NASDAQ
5600 COLUMBIA PIKE
BAILEY'S XROADS VA 22041
703/379-4500

CERMETEK MICROELECTRONICS, INC.
CRMK NASDAQ
1308 BORREGAS AVENUE
SUNNYVALE CA 94089
408/734-8150

CETEC CORPORATION
CEC ASE
9900 BALDWIN PLACE
EL MONTE CA 91731
213/442-8840

CETUS CORPORATION
CTUS NASDAQ
1400 FIFTY-THIRD STREET
EMERYVILLE CA 94608
415/549-3300

CGA COMPUTER ASSOCIATES, INC.
CGAC NASDAQ
255 ROUTE 520
MARLBORO NJ 07746
201/946-8900

CHAMPION SPARK PLUG COMPANY
CHM NYSE
900 UPTON AVENUE
TOLEDO OH 43661
419/535-2567

CHECKPOINT SYSTEMS, INC.
CHEK NASDAQ
P.O. BOX 188
THOROFARE NJ 08086
609/848-1800

CHEM-TECHNICS, INCORPORATED
DTOX NASDAQ
1790 CENTURY BLVD.
ATLANTA GA 30345
404/633-2696

CHEMFIX TECHNOLOGIES, INC.
CFIX NASDAQ
1675 AIRLINE HIGHWAY
KENNER LA 70062
504/467-2800

CHERRY ELECTRICAL CORPORATION
CHER NASDAQ
3600 SUNSET AVENUE
WAUKEGAN IL 60087
312/662-9200

CHIRON CORPORATION
CIRN NASDAQ
4560 HORTON STREET
EMERYVILLE CA 94608
415/655-8730

CHOMERICS, INCORPORATED
CHOM NASDAQ
77 DRAGON COURT
WOBURN MA 01888
617/935-4850

CHRONAR CORPORATION
CRNR NASDAQ
P.O. BOX 177
PRINCETON NJ 08540
609/587-8000

CHYRON CORPORATION
CHYC NASDAQ
265 SPAGNOLI ROAD
MELVILLE NY 11747
516/694-7137

CINCINNATI MILACRON, INC.
CMZ NYSE
4701 MARBURG AVENUE
CINCINNATI OH 45209
513/841-8100

CIPHER DATA PRODUCTS, INC.
CIFR NASDAQ
P.O. BOX 85170
SAN DIEGO CA 92138
714/578-9100

CIRCADIAN, INCORPORATED
CKDN NASDAQ
3960 N. FIRST STREET
SAN JOSE CA 95134
408/943-9222

CIRCON CORPORATION
CCON NASDAQ
749 WARD DRIVE
SANTA BARBARA CA 93111
805/967-0404

CLINI-THERM CORPORATION
CLI NASDAQ
11410 PAGEMILL RD
DALLAS TX 75243
214/343-2180

CLINICAL DATA, INCORPORATED
CLDA NASDAQ
1172 COMMONWEALTH AVENUE
BOSTON MA 02134
617/734-3700

CME-SAT, INCORPORATED
CSAT NASDAQ
444 GULF OF MEXICO DR
LONGBOAT KEY FL 33548
813/383-6447

COBE LABORATORIES, INCORPORATED
COBE NASDAQ
1201 OAK STREET
LAKEWOOD CO 80215
303/232-6800

CODENOLL TECHNOLOGY CORPORATION
CODN NASDAQ
1086 NORTH BROADWAY
YONKERS NY 10701
914/965-6300

COGENIC ENERGY SYSTEMS, INC.
CESI NASDAQ
645 FIFTH AVENUE
NEW YORK NY 10002
212/832-6767

COGNITRONICS CORPORATION
COGN NASDAQ
25 CRESCENT STREET
STAMFORD CT 06906
203/327-5307

COHERENT, INCORPORATED
COHR NASDAQ
3210 PORTER DRIVE
PALO ALTO CA 94303
415/493-2111

COHU, INCORPORATED
COH ASE
P.O. BOX 623
SAN DIEGO CA 92112
714/277-6700

COLECO INDUSTRIES, INCORPORATED
CLO NYSE
945 ASYLUM AVENUE
HARTFORD CT 06105
203/278-0280

COLLABORATIVE RESEARCH, INC.
CRIC NASDAQ
128 SPRING STREET
LEXINGTON MA 02173
617/861-9700

COLLAGEN CORPORATION
CGEN NASDAQ
2455 FABER PLACE
PALO ALTO CA 94303
415/856-0200

COLUMBIA DATA PRODUCTS, INC.
CDPI NASDAQ
9150 RUMSEY ROAD
COLUMBIA MD 21045
301/992-3400

COM SYSTEMS, INCORPORATED
CMSI NASDAQ
157 16 STAGG STREET
VAN NUYS CA 91406
213/988-3140

COMDATA NETWORK, INCORPORATED
CASH NASDAQ
2209 OUSTMOOR ROAD
NASHVILLE TN 37215
615/385-0400

COMDISCO, INCORPORATED
CDO NYSE
6400 SHAFER COURT
ROSEMONT IL 60018
312/698-3000

COMMODORE INTERNATIONAL LIMITED
CBU NYSE
950 RITTENHOUSE ROAD
NORRISTOWN PA 19403
215/666-7950

COMMUNICATIONS SYSTEMS, INC.
CSII NASDAQ
BOX 777
HECTOR MN 55342
612/848-6231

COMMUNICATIONS SATELLITE CORP.
CQ NYSE
950 L'ENFANT PLAZA SW
WASHINGTON DC 20024
202/863-6000

COMMUNICATIONS INDUSTRIES
COMM NASDAQ
1100 FRITO LAY TOWER
DALLAS TX 75235
214/357-4001

COMMUNICATIONS CORP. OF AMERICA
CCPA NASDAQ
8585 N. STEMMONS FRWY
DALLAS TX 75247
214/638-5444

COMP-U-CARD INTER'L, INC.
CUCD NASDAQ
777 SUMMER STREET
STAMFORD CT 06901
203/324-9261

COMPACT VIDEO, INCORPORATED
CVSI NASDAQ
2813 W. ALAMEDA BLVD
BURBANK CA 91505
213/840-7000

COMPAQ COMPUTER CORPORATION
CMPQ NASDAQ
20333 FM 149
HOUSTON TX 77070
713/370-7040

COMPRESSION LABS, INC.
CLIX NASDAQ
2305 BERING DRIVE
SAN JOSE CA 95131
408/946-3060

COMPTEK RESEARCH, INCORPORATED
CMTK NASDAQ
45 OAK STREET
BUFFALO NY 14203
716/842-2700

COMPUCARE, INCORPORATED
CMPC NASDAQ
8200 GREENSBORO DRIVE
MCLEAN VA 22102
703/821-8858

COMPUCORP
CCUP NASDAQ
2211 MICHIGAN AVENUE
SANTA MONICA CA 90404
213/829-7453

COMPUGRAPHIC CORPORATION
CPU NYSE
200 BALLARDVALE STREET
WILMINGTON MA 01887
617/944-6555

COMPUSCAN, INCORPORATED
CSCN NASDAQ
900 HUYLER STREET
TETERBORO NJ 07608
201/575-0500

COMPUSHOP, INCORPORATED
CSHP NASDAQ
1355 GLENVILLE DRIVE
RICHARDSON TX 75081
214/783-1252

COMPUTER & COMMUNICATION TECH.
CCTC NASDAQ
495 S. FAIRVIEW AVENUE
SANTA BARBARA CA 93117
805/964-0771

COMPUTER ASSOC. INTER'L, INC.
CASI NASDAQ
125 JERICHO TURNPIKE
JERICHO NY 11753
516/333-6700

COMPUTER ASSOCIATES INT'L, INC.
CASI NASDAQ
125 JERICHO TURNPIKE
JERICHO NY 11753
516/333-6700

COMPUTER AUTOMATION, INC.
CAUT NASDAQ
4890 STERLING DRIVE
BOULDER CO 80301
303/444-8748

COMPUTER DEPOT, INC.
CDPT NASDAQ
7464 WEST 78TH ST.
EDINA MN 55435
612/944-8780

COMPUTER DOCTOR INCORPORATED
CDRX NASDAQ
485 WASHINGTON AVENUE
PLEASANTVILLE NY 10604
914/747-2777

COMPUTER DIALYSIS SYSTEMS, INC.
CODI NASDAQ
2400 CENTRAL AVENUE
BOULDER CO 80301
303/443-3131

COMPUTER DESIGNED SYSTEMS, INC.
DCSI NASDAQ
10911 OLSON MEMORIAL HWY
MINNEAPOLIS MN 55441
612/545-2855

COMPUTER DEVICES, INCORPORATED
CDIT NASDAQ
25 NORTH AVENUE
BURLINGTON MA 01803
617/273-1550

COMPUTER DATA SYSTEMS, INC.
CPTD NASDAQ
7315 WISCONSIN AVENUE
BETHESDA MD 20814
301/657-1730

COMPUTER ENTRY SYSTEMS CORP.
CESC NASDAQ
2141 INDUSTRIAL PARKWAY
SILVER SPRING MD 20904
301/622-3500

COMPUTER FACTORY, INC. (THE)
CFA ASE
485 LEXINGTON AVENUE
NEW YORK NY 10017
212/687-5000

COMPUTER HORIZONS CORPORATION
CHRZ NASDAQ
747 THIRD AVENUE
NEW YORK NY 10017
212/371-9600

COMPUTER IDENTICS CORPORATION
CIDN NASDAQ
5 SHAWMUT ROAD
CANTON MA 02021
617/821-0830

COMPUTER LANGUAGE RESEARCH, INC
CLRI NASDAQ
2395 MIDWAY ROAD
CARROLLTON TX 75006
214/934-7000

COMPUTER MICROFILM CORPORATION
COMI NASDAQ
1699 TULLIE CIRCLE
ATLANTA GA 30329
404/321-0886

COMPUTER NETWORK CORPORATION
CNET NASDAQ
5185 MACARTHUR BLVD NW
WASHINGTON DC 20016
202/537-2500

COMPUTER PRODUCTS, INCORPORATED
CPRD NASDAQ
1400 N.W. 70TH STREET
FT. LAUDERDALE FL 33309
305/974-5500

COMPUTER RESOURCES, INC.
CRII NASDAQ
4520 W 160TH STREET
CLEVELAND OH 44135
216/362-1020

COMPUTER RESEARCH, INCORPORATED
CORE NASDAQ
P.O. BOX 325
PITTSBURGH PA 15230
412/262-4430

COMPUTER STORE (THE), INC.
TCSI NASDAQ
56 UNION AVENUE
SUDBURY MA 01776
617/879-3700

COMPUTER SYNERGY, INCORPORATED
CSYN NASDAQ
2201 BROADWAY
OAKLAND CA 94612
415/444-3434

COMPUTER SCIENCES CORPORATION
CSC NYSE
2100 EAST GRAND AVE.
EL SEGUNDO CA 90245
213/615-0311

COMPUTER TRANSCEIVER SYS., INC.
CTRC NASDAQ
EAST 66 MIDLAND AVENUE
PARAMUS NJ 07652
201/261-6800

COMPUTER TERMINAL SYSTEMS, INC.
CTML NASDAQ
89 ARKAY DRIVE
HAUPPAGUE NY 11788
516/435-2900

COMPUTER TASK GROUP, INC.
CTSK NASDAQ
800 DELAWARE AVENUE
BUFFALO NY 14209
716/882-8000

COMPUTER USAGE COMPANY
CUSE NASDAQ
150 FOURTH STREET
SAN FRANCISCO CA 94103
415/543-3940

COMPUTERCRAFT, INCORPORATED
CRFT NASDAQ
1616 SOUTH VOSS ROAD, STE 900
HOUSTON TX 77057
713/977-8419

COMPUTERS FOR LESS, INCORPORATED
CFLI NASDAQ
50 MARKET BLVD
HAUPPAUGE NY 11788
516/435-8099

COMPUTERVISION CORPORATION
CVN NYSE
201 BURLINGTON ROAD
BEDFORD MA 01730
617/275-1800

COMPUTONE SYSTEMS, INCORPORATED
CTON NASDAQ
1 DUNWOODY PARK
ATLANTA GA 30338
404/393-3010

COMSERV CORPORATION
CMSV NASDAQ
3400 COMSERV DRIVE
EAGAN MN 55122
612/452-7770

COMSHARE, INCORPORATED
CSRE NASDAQ
3001 S. STATE STREET
ANN ARBOR MI 48104
313/994-4800

COMTECH TELECOMM. CORPORATION
CMTL NASDAQ
45 OSER AVENUE
HAUPPAUGE NY 11788
516/496-7049

COMTEX SCIENTIFIC CORPORATION
CMTX NASDAQ
850 THIRD AVENUE
NEW YORK NY 10022
212/838-7200

CONCORD COMPUTING CORPORATION
CEFT NASDAQ
7 ALFRED CIRCLE
BEDFORD MA 01730
617/275-1730

CONOLOG CORPORATION
CNLG NASDAQ
5 COLUMBIA ROAD
SOMERVILLE NJ 08686
201/722-3770

CONRAC CORPORATION
CAX NYSE
THREE LANDMARK SQUARE
STAMFORD CT 06901
203/348-2100

CONTINUUM COMPANY, INC. (THE)
CTUC NASDAQ
3429 EXECUTIVE CENTER DRIVE
AUSTIN TX 78731
512/345-5700

CONTROL DATA CORPORATION
CDA NYSE
8100 34TH AVE, SOUTH
MINNEAPOLIS MN 55420
612/853-8100

CONTROL LASER CORPORATION
CLSR NASDAQ
11222 ASTRONAUT BOULEVARD
ORLANDO FL 32809
305/851-2540

CONVERGENT TECHNOLOGIES, INC.
CVGT NASDAQ
2500 AUGUSTINE DRIVE
SANTA CLARA CA 95051
408/727-8830

COOPER BIOMEDICAL, INCORPORATED
BUGS NASDAQ
3145 PORTER DRIVE
PALO ALTO CA 94304
415/856-5000

COOPER LABORATORIES, INC.
COO NYSE
3145 PORTER DRIVE
PALO ALTO CA 94304
415/856-5000

COPPERWELD CORPORATION
COS NYSE
TWO OLIVER PLAZA
PITTSBURG PA 15222
412/263-3200

COPYTELE, INCORPORATED
COPY NASDAQ
900 WALT WHITMAN RD
HUNTINGTON STATION NY 11746
516/549-5900

CORCOM, INCORPORATED
CORC NASDAQ
1600 WINCHESTER ROAD
LIBERTYVILLE IL 60048
312/680-7400

CORDIS CORPORATION
CORD NASDAQ
P.O. BOX 525700
MIAMI FL 33152
305/551-2000

CORVUS SYSTEMS, INCORPORATED
CRVS NASDAQ
2029 O'TOOLE AVENUE
SAN JOSE CA 95131
408/946-7700

CPT CORPORATION
CPTC NASDAQ
8100 MITCHELL ROAD
EDEN PRARIE MN 55440
612/937-8000

CRAY RESEARCH, INCORPORATED
CYR NYSE
1440 NORTHLAND DRIVE
MENDOTA HEIGHTS MN 55120
612/452-6650

CREATIVE COMPUTER APP., INC.
CCAI NASDAQ
23961 CRAFTSMAN ROAD
CALABASAS CA 91302
818/348-8089

CRIME CONTROL, INCORPORATED
CRIM NASDAQ
3660 N. WASHINGTON BLVD
INDIANAPOLIS IN 46205
317/247-7770

CROSS & TRECKER CORPORATION
CTCO NASDAQ
P.O. BOX 925
BLOOMFIELD HILLS MI 48013
313/644-4343

CSP, INCORPORATED (CSPI)
CSPI NASDAQ
40 LINNELL CIRCLE
BILLERICA MA 01821
617/272-6020

CUBIC CORPORATION
CUB ASE
P.O. BOX 80787
SAN DIEGO CA 92138
714/277-6780

CULLINET SOFTWARE, INCORPORATED
CUL NYSE
400 BLUE HILL DRIVE
WESTWOOD MA 02090
617/329-7700

CURTISS-WRIGHT CORPORATION
CW NYSE
ONE PASSAIC STREET
WOOD-RIDGE NJ 07075
201/777-2900

CYBER DIAGNOSTICS (CDX), INC.
CDXX NASDAQ
10691 E. BETHANY DRIVE
AURORA CO 80014
303/695-8751

CYBERTEK COMPUTER PRODUCTS
CKCP NASDAQ
6133 BRISTOL PARKWAY
CULVER CITY CA 90230
213/649-2450

CYCARE SYSTEMS, INCORPORATED
CYCR NASDAQ
P.O. BOX 1278
DUBUQUE IA 52001
319/556-3131

CYCLOTRON CORPORATION (THE)
CYCC NASDAQ
950 GILMAN STREET
BERKELEY CA 94710
415/524-8670

CYTOGEN CORPORATION
CYTO NASDAQ
201 COLLEGE RD E./FORRESTAL CTR
PRINCETON NY 08540
609/452-8838

CYTOX CORPORATION
CYTX NASDAQ
605 MADISON AVENUE
NEW YORK NY 10022
212/486-9120

DAISY SYSTEMS CORPORATION
DAZY NASDAQ
139 KIFER COURT
SUNNYVALE CA 94086
408/773-9111

DAMON CORPORATION
DMN NYSE
115 FOURTH AVENUE
NEEDHAM HTS MA 02194
617/449-0800

DATA ARCHITECTS, INCORPORATED
DRCH NASDAQ
245 WINTER STREET
WALTHAM MA 02154
617/890-7730

DATA CARD CORPORATION
DATC NASDAQ
P.O. BOX 9355
MINNEAPOLIS MN 55440
612/933-1223

DATA GENERAL CORPORATION
DGN NYSE
4400 COMPUTER DRIVE
WESTBORO MS 01580
617/366-8911

DATA I/O CORPORATION
DAIO NASDAQ
10525 WILLOWS ROAD, N.E.
REDMOND WA 98052
206/881-6444

DATA PACKAGING CORPORATION
DPKG NASDAQ
205 BROADWAY
CAMBRIDGE MA 02139
617/868-6200

DATA SWITCH CORPORATION
DASW NASDAQ
444 WESTPORT AVENUE
NORWALK CT 06851
203/847-9800

DATA TRANSLATION, INC.
DATX NASDAQ
100 LOCKE DRIVE
MARLBORO MA 01752
617/481-3700

DATA-DESIGN LABORATORIES
DDES NASDAQ
7925 CENTER AVENUE
CUCAMONGA CA 91730
714/987-2511

DATACOPY CORPORATION
DCPY NASDAQ
170 EAST MEADOW CIRCLE
PALO ALTO CA 94303
415/965-7900

DATAKEY, INCORPORATED
DKEY NASDAQ
12281 NICOLLET AVENUE
BURNSVILLE MN 55337
612/890-6850

DATAPOINT CORPORATION
DPT NYSE
7900 CALLAGHAN ROAD
SAN ANTONIO TX 78229
512/699-4478

DATAPOWER, INCORPORATED
DPWR NASDAQ
3328 WEST FIRST STREET
SANTA ANA CA 92703
714/775-2000

DATAPRODUCTS CORPORATION
DPC ASE
6200 CANOGA AVENUE
WOODLAND HILLS CA 91365
213/887-8000

DATARAM CORPORATION
DTM ASE
PRINCETON ROAD
CRANBURY NJ 08512
609/799-0071

DATASCOPE CORPORATION
DSCP NASDAQ
580 WINTERS AVENUE
PARAMUS NJ 07652
201/265-8800

DATASOUTH COMPUTER CORPORATION
DSCC NASDAQ
4216 STUART ANDREW BLVD
CHARLOTTE NC 28210
704/523-8500

DATATRAK, INCORPORATED
DTRK NASDAQ
1700 STIERLIN ROAD
MOUNTAIN VIEW CA 94043
415/967-3911

DATATRON, INCORPORATED
DTRN NASDAQ
2942 DOW AVENUE
TUSTIN CA 92680
714/544-9970

DATAVISION, INCORPORATED
DVIS NASDAQ
16545 EASTLAND
ROSEVILLE MI 48066
313/775-7077

DATRICON CORPORATION
DATN NASDAQ
155 B AVENUE
LAKE OSWEGO OR 97034
503/636-7671

DATUM, INCORPORATED
DATM NASDAQ
1363 S STATE COLLEGE BLVD
ANAHEIM CA 92806
714/533-6333

DAWSON GEOPHYSICAL COMPANY
DWSN NASDAQ
208 S. MARIENFELD STREET
MIDLAND TX 79701
915/682-7356

DBA SYSTEMS, INCORPORATED
DBAS NASDAQ
1103 W. HIBISCUS BLVD
MELBOURNE FL 32901
305/725-3711

DECISION DATA COMPUTER CORP.
DDCC NASDAQ
100 WITMER ROAD
HORSHAM PA 19044
215/674-3300

DECISION SYSTEMS, INCORPORATED
DCSN NASDAQ
200 ROUTE 17
MATWAH NJ 07430
201/529-1440

DECOM SYSTEMS, INCORPORATED
DSII NASDAQ
340 RANCHEROS DRIVE
SAN MARCOS CA 92069
619/744-1002

DEKALB AGRESEARCH
DKLBB NASDAQ
SYCAMORE ROAD
DEKALB IL 60115
815/756-3671

DELMED, INCORPORATED
DMED NASDAQ
437 TURNPIKE, SOUTH
CANTON MA 02021
617/821-0500

DELTA DATA SYSTEMS CORPORATION
DDSC NASDAQ
2595 METROPOLITAN DR
TREVOSE PA 19047
215/322-5400

DELTAX CORPORATION
DLTK NASDAQ
13330 12TH AVENUE N.
MINNEAPOLIS MN 55441
612/544-3371

DENELCOR, INCORPORATED
DENL NASDAQ
1700 E. OHIO PLACE
AURORA CO 80017
303/337-7900

DENNING MOBILE ROBOTICS, INC.
GARD NASDAQ
21 CUMMINGS PARK
WOBURN MA 01801
617/935-4840

DETECTION SYSTEMS, INCORPORATED
DETC NASDAQ
400 MASON ROAD
FAIRPORT NY 14450
716/223-4060

DETECTOR ELECTRONICS CORP.
DETX NASDAQ
6901 W. 110 STREET
MINNEAPOLIS MN 55438
612/941-5665

DEVRY, INCORPORATED
DVRY NASDAQ
2201 WEST HOWARD STREET
EVANSTON IL 60202
312/328-8100

DEWEY ELECTRONICS CORPORATION
DEWY NASDAQ
27 MULLER ROAD
OAKLAND NJ 07436
201/337-4700

DH TECHNOLOGY, INCORPORATED
DHTK NASDAQ
754 NORTH PASTORIA AVENUE
SUNNYVALE CA 94086
408/738-2082

DIAGNOSTIC PRODUCTS CORPORATION
DPCZ NASDAQ
5700 W. 96TH STREET
LOS ANGELES CA 90045
213/776-0180

DIAGNOSTIC/RETRIEVAL SYS., INC.
DRSI NASDAQ
16 THORNTON ROAD
OAKLAND NJ 07436
201/337-3800

DIASONICS, INCORPORATED
DNIC NASDAQ
1708 MCCARTHY BLVD
MILIPITAS CA 95035
408/946-9001

DICEON ELECTRONICS, INC.
DICN NASDAQ
2500 MICHELSON DRIVE
IRVINE CA 92714
714/833-0870

DICOMED CORPORATION
DCOM NASDAQ
9700 NEWTON AVE, S.
MINNEAPOLIS MN 55431
612/887-7100

DIGILOG, INCORPORATED
DILO NASDAQ
1370 WELSH ROAD
MONTGOMERYVILLE PA 18936
215/628-4530

DIGITAL COMM. ASSOCIATES, INC.
DCAI NASDAQ
303 TECHNOLOGY PARK
NORCROSS GA 30092
404/448-1400

DIGITAL DATACOM, INCORPORATED
DDII NASDAQ
27721 SOUTH LA PAZ RD
LAGUNA NIGUEL CA 92677
714/831-8470

DIGITAL EQUIPMENT CORP.
DEC NYSE
146 MAIN STREET
MAYNARD MA 01754
617/493-5523

DIGITAL PRODUCTS CORPORATION
DIPC NASDAQ
4021 N.E. 5TH TERRACE
FORT LAUDERDALE FL 33334
305/564-0521

DIGITAL SWITCH CORPORATION
DIGI NASDAQ
P.O. BOX 911
RICHARDSON TX 75080
214/234-3000

DIGITAL SYSTEMS CORPORATION
DIGS NASDAQ
3 NORTH MAIN STREET
WALKERSVILLE MD 21793
301/845-4141

DIMIS INCORPORATED
DMIS NASDAQ
1806 ROUTE 35
OCEAN NJ 07712
201/671-1011

DIONEX CORPORATION
DNEX NASDAQ
1228 TITAN WAY
SUNNYVALE CA 94086
408/737-0700

DIPLOMAT ELECTRONICS CORP.
DPLT NASDAQ
110 MARCUS DRIVE
MELVILLE NY 11747
516/454-6334

DISTRIBUTED LOGIC CORPORATION
DLOG NASDAQ
P.O. BOX 6270
ANAHEIM CA 92806
714/937-5700

DKM ELECTRONICS, INCORPORATED
DKME NASDAQ
100 ALEXANDER AVENUE
POMPTON PLAINS NJ 07444
201/835-6785

DNA MEDICAL, INCORPORATED
DNAM NASDAQ
3385 WEST 1820 SOUTH
SALT LAKE CITY UT 84104
801/973-4600

DNA PLANT TECHNOLOGY CORP.
DNAP NASDAQ
2611 BRANCH PIKE
CINNAMINSON NJ 08077
609/829-0110

DOCUTEL/OLIVETTI CORPORATION
DCTL NASDAQ
106 DECKER COURT
IRVING TX 75062
214/258-8610

DOTRONIX, INCORPORATED
DOTX NASDAQ
160 FIRST STREET S.E.
NEW BRIGHTON MN 55112
612/633-1742

DRANETZ ENGINEERING LABS, INC.
DRAN NASDAQ
1000 NEW DURHAM ROAD
EDISON NJ 08818
201/287-3680

DREXLER TECHNOLOGY CORP.
DRXR NASDAQ
3960 FABIAN WAY
PALO ALTO CA 94303
415/969-7277

DST SYSTEMS, INCORPORATED
DSTS NASDAQ
301 WEST 11TH STREET
KANSAS CITY MO 64105
816/421-4343

DUCOMMUN INCORPORATED
DCO ASE
612 S. FLOWER STREET STE 460
LOS ANGELES CA 90017
213/612-4200

DUQUESNE SYSTEMS, INC.
DUQN NASDAQ
TWO ALLEGENY CENTER
PITTSBURG PA 15212
412/323-2600

DURALITH CORPORATION
DRLH NASDAQ
525 ORANGE STREET
MILLVILLE NJ 08332
609/825-6900

DYATRON CORPORATION
DYTR NASDAQ
P.O. BOX 235
BIRMINGHAM AL 35201
205/956-7500

DYNALECTRON CORPORATION
DYN ASE
1313 DOLLEY MADISON BOULEVARD
MCLEAN VA 22101
703/356-0480

DYNAMICS RESEARCH CORPORATION
DRCO NASDAQ
60 CONCORD STREET
WILMINGTON MA 01887
617/658-6100

DYNASCAN CORPORATION
DYNA NASDAQ
6460 WEST CORTLAND STREET
CHICAGO IL 60635
312/889-8870

DYNATECH CORPORATION
DYTC NASDAQ
3 NEW ENGLAND EXC. PK.
BURLINGTON MA 01803
617/272-3304

DYRATRON CORPORATION
DYTR NASDAQ
P.O. BOX 235
BIRMINGHAM AL 35201
205/956-7500

DYSAN CORPORATION
DYSN NASDAQ
5440 PATRICK HENRY DRIVE
SANTA CLARA CA 95050
408/988-3472

E-H INTERNATIONAL, INCORPORATED
EHIL NASDAQ
696 EAST TRIMBLE ROAD
SAN JOSE CA 95131
408/946-9100

E-SYSTEMS, INCORPORATED
ESY NYSE
6250 LBJ FREEWAY
DALLAS TX 75266
214/661-1000

EASTMAN KODAK COMPANY
EK NYSE
343 STATE STREET
ROCHESTER NY 14650
716/724-4000

EATON CORPORATION
ETN NYSE
100 ERIEVIEW PLAZA
CLEVELAND OH 44114
216/523-5000

EDO CORPORATION
EDO ASE
14-04 111TH STREET
COLLEGE POINT NY 11356
212/445-6000

EDUCATIONAL DEVELOPMENT CORP.
EDUC NASDAQ
8141 EAST 44TH STREET
TULSA OK 74145
918/622-4522

EDUCATIONAL TECHNOLOGY, INC.
ETEC NASDAQ
2224 HEWLETT AVENUE
MERRICK NY 11566
516/623-3200

EDUCATIONAL COMPUTER CORP.
EDCC NASDAQ
5882 SOUTH TAMPA AVENUE
ORLANDO FL 32809
305/859-7410

EDUCOM CORPORATION
EDCN NASDAQ
210 W. WASHINGTON SQ.
PHILADELPHIA PA 19106
215/238-5250

EG&G, INCORPORATED
EGG NYSE
45 WILLIAMS STREET
WELLESLEY MA 02181
617/237-5100

EIKONIX CORPORATION
KONX NASDAQ
23 CROSBY DRIVE
BEDFORD MA 01730
617/275-5070

EIP MICROWAVE, INCORPORATED
EIPM NASDAQ
4500 CAMPUS DRIVE
NEWPORT BEACH CA 92660
714/540-6655

ELECTRO-BIOLOGY, INCORPORATED
EBII NASDAQ
300 FAIRFIELD ROAD
FAIRFIELD NJ 07006
201/575-9201

ELECTRO-CATHETER CORPORATION
ECTH NASDAQ
2100 FELVER COURT
RAHWAY NJ 07065
201/382-5600

ELECTRO-NUCLEONICS, INC.
ENUC NASDAQ
P.O. BOX 803
FAIRFIELD NJ 07006
201/227-6700

ELECTRO-SCIENTIFIC INDUSTRIES
ESIO NASDAQ
13900 NW SCIENCE PARK DRIVE
PORTLAND OR 97229
503/641-4141

ELECTRO-SENSORS, INCORPORATED
ELSE NASDAQ
7301 WASHINGTON AVENUE S
MINNEAPOLIS MN 55435
612/941-8171

ELECTROMAGNETIC SCIENCES, INC.
ELMG NASDAQ
125 TECHNOLOGY PARK/ATLANTA
NORCROSS GA 30092
404/448-5770

ELECTROMEDICS, INCORPORATED
ELMD NASDAQ
109 INVERNESS DRIVE EAST
ENGELWOOD CO 80112
303/770-8704

ELECTRONIC MAIL CORP. OF AM.
EMCA NASDAQ
30 ROCKEFELLER PLAZA STE 4310
NEW YORK NY 10112
212/956-4700

ELECTRONICS CORP. OF AMERICA
ECA ASE
ONE MEMORIAL DRIVE
CAMBRIDGE MA 02142
617/864-8000

ELECTRONIC MEMORIES & MAGNETICS
EMM NYSE
16000 VENUTRA BLVD SUITE 801
ENCINO CA 94436
213/995-1755

ELECTRONIC DATA SYSTEMS CORP.
EDS NYSE
771 FOREST LANE
DALLAS TX 75230
214/661-6000

ELECTRONIC MODULES CORPORATION
EDOM NASDAQ
P.O. BOX 141
TIMONIUM MD 21093
301/667-4800

ELECTRONIC ASSOCIATES, INC.
EA NYSE
185 MONMOUTH PARKWAY
WEST LONG BRANCH NJ 07764
201/229-1100

ELECTROSPACE SYSTEMS, INC.
ELEC NASDAQ
1601 N. PLANO ROAD
RICHARDSON TX 75081
214/231-9303

EMERSON RADIO CORPORATION
EME NYSE
1 EMERSON LANE
SECAUCUS NJ 07094
201/865-4343

EMULEX CORPORATION
EMLX NASDAQ
2001 E. DEERE AVENUE
SANTA ANA CA 92705
714/557-7580

ENDATA, INCORPORATED
DATA NASDAQ
50 VANTAGE WAY
NASHVILLE TN 37228
615/244-0244

ENDOTRONICS, INCORPORATED
ENDO NASDAQ
P.O. BOX 32366
MINNEAPOLIS MN 55432
612/786-0302

ENERGY CONVERSION DEVICES, INC.
ENER NASDAQ
1675 WEST MAPLE ROAD
TROY MI 48084
313/280-1900

ENERGY FACTORS, INCORPORATED
EFAC NASDAQ
1495 PACIFIC HIGHWAY, STE 400
SAN DIEGO CA 92101
619/239-9900

ENERGY OPTICS, INCORPORATED
EOPT NASDAQ
224 NORTH CAMPO STREET
LAS CRUCES NM 88001
505/523-4561

ENGINEERED SYS. & DEV. CORP.
ESD AMEX
600 MERIDIAN AVENUE
SAN JOSE CA 95126
408/280-5000

ENGINEERING MEASUREMENTS CO.
EMCO NASDAQ
600 DIAGONAL HIGHWAY
LONGMONT CO 80501
303/651-0550

ENTRE COMPUTER CENTERS
ETRE NASDAQ
1951 KIDWELL DRIVE
VIENNA VA 22180
703/556-0800

ENVIRONMENTAL TECHNOLOGY, INC.
ETUS NASDAQ
1018 28TH STREET
ORLANDO FL 32805
305/843-0400

ENVIRONMENTAL TECTONICS CORP.
ENVT NASDAQ
COUNTY LINE INDUSTRIAL PARK
SOUTHHAMPTON PA 18966
215/355-9100

ENVIRONMENTAL SYSTEMS COMPANY
ESCO NASDAQ
1015 LOUISIANA STREET
LITTLE ROCK AR 72202
501/376-8142

ENVIRONMENTAL TESTING & CERTIFICATION
ETCC NASDAQ
284 RARITAN CENTER PARKWAY
EDISON NJ 08837
201/225-5600

ENZO BIOCHEM, INCORPORATED
ENZO NASDAQ
325 HUDSON STREET
NEW YORK NY 10013
212/741-3838

EPSILON DATA MANAGEMENT, INC.
EPSI NASDAQ
1 NEW ENGLAND EXEC. PK.
BURLINGTON MA 01803
617/273-0250

EQUATORIAL COMMUNICATIONS CO.
EQUA NASDAQ
300 FERGUSON DRIVE
MOUNTAIN VIEW CA 94043
415/969-9500

EQUINOX SOLAR, INCORPORATED
EQIX NASDAQ
3350 N.W. 60TH STREET
MIAMI FL 33142
305/638-9500

ESQUIRE, INCORPORATED
ESQ NYSE
488 MADISON AVENUE
NEW YORK NY 10022
212/407-0300

ESSEX CORPORATION
ESEX NASDAQ
333 NORTH FAIRFAX STREET
ALEXANDRIA VA 22314
703/548-4500

ESTERLINE CORPORATION
ESL NYSE
1120 POST ROAD
DARIEN CT 06820
203/655-7651

EVALUATION RESEARCH (ERC) CORP.
ERC ASE
2070 CHAIN BRIDGE ROAD, STE 400
VIENNA VA 22180
703/827-0720

EVANS & SUTHERLAND COMPUTER CO.
ESCC NASDAQ
580 ARAPEEN DRIVE
SALT LAKE CITY UT 84108
801/582-5847

FAFCO, INCORPORATED
FAFO NASDAQ
255 CONSTITUTION DRIVE
MENLO PARK CA 94025
415/321-3650

FAIRCHILD INDUSTRIES, INC.
FEN NYSE
20301 CENTURY BLVD
GERMANTOWN MD 20874
301/428-6000

FARED ROBOT SYSTEMS, INC.
FARE NASDAQ
3860 REVERE STREET
DENVER CO 80239
303/371-5868

FDP CORPORATION
FDPC NASDAQ
2675 SOUTH BAYSHORE DRIVE
MIAMI FL 33133
305/858-8200

FEDERATED GROUP, INC. (THE)
FEGP NASDAQ
5655 E. UNION PACIFIC AVENUE
CITY - COMMERCE CA 90022
213/728-5100

FERROFLUIDICS CORPORATION
FERO NASDAQ
40 SIMON STREET
NASHUA NH 03061
603/883-9800

FIBRONICS INTERNATIONAL, INC.
FBRX NASDAQ
218 WEST MAIN STREET
HYANNIS MA 02601
617/778-0700

FINGERMATRIX, INC.
FINX NASDAQ
30 VIRGINIA ROAD
N. WHITE PLAINS NY 10603
914/428-5441

FINNIGAN CORPORATION
FNNG NASDAQ
355 RIVER OAKS P'KWAY
SAN JOSE CA 95134
408/946-4848

FIRST FINANCIAL MGMT. CORP.
FFMC NASDAQ
2695 BUFORD HIGHWAY, N.E.
ATLANTA GA 30324
404/325-9715

FIRST MIDWEST CORPORATION
FMWC NASDAQ
15 SOUTH FIFTH STREET
MINNEAPOLIS MN 55402
612/339-9391

FISCHER & PORTER COMPANY
FP ASE
COUNTY LINE ROAD
WARMINSTER PA 18974
215/674-6000

FLEXIBLE COMPUTER CORPORATION
FLXXA NASDAQ
1801 ROYAL LANE, STE 810
DALLAS TX 75229
214/869-1234

FLIGHT DYNAMICS, INCORPORATED
FLYT NASDAQ
P.O. BOX 1079
HILLSBORO OR 97123
503/640-8955

FLOATING POINT SYSTEMS, INC.
FLP NYSE
P.O. BOX 23489
PORTLAND OR 97005
503/641-3151

FLOW GENERAL, INCORPORATED
FGN NYSE
7655 OLD SPRINGHOUSE ROAD
MCLEAN VA 22102
703/893-5915

FLOW SYSTEMS, INCORPORATED
FLOW NASDAQ
21440 68TH AVENUE SOUTH
KENT WA 98032
206/938-3569

FLUKE (JOHN) MFG. CO., INC.
FKM ASE
6920 SEAWAY BLVD
EVERETT WA 98206
206/356-5310

FONAR CORPORATION
FONR NASDAQ
110 MARCUS DRIVE
MELVILLE NY 11746
516/694-2929

FORMASTER CORPORATION
FMSR NASDAQ
1983 CONCOURSE DRIVE
SAN JOSE CA 95131
408/942-1771

FORTUNE SYSTEMS CORPORATION
FSYS NASDAQ
300 HARBOR BLVD.
BELMONT CA 94002
415/593-9000

FOXBORO COMPANY (THE)
FOX NYSE
38 NEPONSET AVENUE
FOXBORO MA 02035
617/543-8750

FREQUENCY ELECTRONICS, INC.
FEI ASE
50 CHARLES LINDBERG BLVD
MITCHEL FIELD NY 11553
516/794-4500

FRIGITRONICS, INCORPORATED
FRG NYSE
770 RIVER ROAD
SHELTON CT 06484
203/929-6321

GALILEO ELECTRO-OPTICS CORP.
GAEO NASDAQ
GALILEO PARK
STURBRIGDE MA 01518
617/347-9191

GAME-A-TRON CORPORATION
GAME NASDAQ
45 OSGOOD AVENUE
NEW BRITAIN CT 06053
203/223-2760

GAMES NETWORK, INC. (THE)
GNET NASDAQ
4401 WILSHIRE BOULEVARD
LOS ANGELES CA 90005
213/932-1950

GAMMA BIOLOGICALS, INCORPORATED
GAMA NASDAQ
3700 MANGUM ROAD
HOUSTON TX 77092
713/681-8481

GCA CORPORATION
GCA NYSE
209 BURLINGTON ROAD
BEDFORD MA 01730
617/275-9000

GELMAN SCIENCES, INCORPORATED
GSC ASE
600 SOUTH WAGNER ROAD BOX 1448
ANN ARBOR MI 48106
313/665-0651

GENENTECH, INCORPORATED
GENE NASDAQ
460 POINT SAN BRUNO BLVD
S. SAN FRANCISCO CA 94080
415/952-1000

GENERAL AUTOMATION, INC.
GENA NASDAQ
P.O. BOX 4883
ANAHEIM CA 92803
714/778-4800

GENERAL CERAMICS, INC.
NBEL NASDAQ
HASKELL NJ 07420
201/839-1600

GENERAL DATACOMM IND., INC.
GDC NYSE
ONE KENNEDY AVENUE
DANBURY CT 06810
203/797-0711

GENERAL DYNAMICS CORPORATION
GD NYSE
PIERRE LACLEDE CENTER
ST. LOUIS MO 63105
314/889-8200

GENERAL ELECTRIC (GE) COMPANY
GE NYSE
3135 EASTERN TURNPIKE
FAIRFIELD CT 06431
203/373-2211

GENERAL GENETICS CORPORATION
GENG NASDAQ
154 WEST 44TH AVENUE
GOLDEN CO 80403
303/279-7349

GENERAL INSTRUMENT CORPORATION
GRL NYSE
767 FIFTH AVENUE
NEW YORK NY 10153
212/207-6200

GENERAL MICROWAVE CORPORATION
GMIC NASDAQ
155 MARINE STREET
FARMINGDALE NY 11735
516/694-3600

GENERAL PHYSICS CORPORATION
GPHY NASDAQ
1000 CENTURY PLAZA
COLUMBIA MD 21044
301/730-4055

GENERAL SIGNAL CORPORATION
GSX NYSE
HIGH RIDGE PARK
STAMFORD CT 06904
203/357-8800

GENETIC ENGINEERING, INC.
GEEN NASDAQ
P.O. BOX 33554
DENVER CO 80233
303/457-1311

GENETIC LABORATORIES, INC.
GEML NASDAQ
1385 CENTENNIAL DRIVE
ST PAUL MN 55113
612/636-4112

GENETIC SYSTEMS CORPORATION
GENS NASDAQ
3005 FIRST AVENUE
SEATTLE WA 98121
206/624-4300

GENEX CORPORATION
GNEX NASDAQ
6110 EXECUTIVE BLVD
ROCHVILLE MD 20852
301/770-0650

GENISCO TECHNOLOGY CORPORATION
GES ASE
18435 SUSANA ROAD
RANCHO DOMINGUEZ CA 90221
213/537-4750

GENRAD, INCORPORATED
GEN NYSE
300 BAKER AVENUE
CONCORD MA 01742
617/369-4400

GENTEX CORPORATION
GNTX NASDAQ
10985 CHICAGO DRIVE
ZEELAND MI 49464
616/392-7195

GENTRONIX LABORATORIES, INC.
PENDING NASDAQ
15825 SHADY GROVE ROAD
ROCKVILLE MD 20850
301/948-7400

GEOTHERMAL RESOURCES INTER'L
GEO ASE
545 MIDDLEFIELD ROAD, STE 200
MENLO PARK CA 94025
415/326-5470

GERBER SCIENTIFIC, INCORPORATED
GRB NYSE
P.O. BOX 305
HARTFORD CT 06101
203/644-1551

GERBER SYSTEMS TECHNOLOGY, INC.
GSTI NASDAQ
40 GERBER ROAD EAST
SOUTH WINDSOR CT 06074
203/644-2581

GIGA-TRONICS, INCORPORATED
GIGA NASDAQ
2495 ESTLAND WAY
PLEASANT HILL CA 94523
415/680-8160

GLOBUSCOPE, INCORPORATED
GPIX NASDAQ
44 W 24TH STREET
NEW YORK NY 10010
212/243-1000

GOULD, INCORPORATED
GLD NYSE
10 GOULD CENTER
ROLLING MEADOWS IL 60008
312/640-4000

GRACO, INCORPORATED
GRAC NASDAQ
60 ELEVENTH AVE, N.E.
MINNEAPOLIS MN 55413
612/623-6000

GRAPHIC SCANNING CORPORATION
GSCC NASDAQ
3298 ALFRED AVENUE
TEANECK NJ 07666
201/837-5100

GREATER WASHINGTON INVESTORS
GWII NASDAQ
5454 WISCONSIN AVENUE
CHEVY CHASE MD 20815
301/656-0626

GRUMMAN CORPORATION
GQ NYSE
1111 STEWART AVENUE
BETHPAGE NY 11714
516/575-3344

GTE CORPORATION
GTE ASE
ONE STAMFORD FORUM
STAMFORD CT 06904
203/965-2000

GTECH CORPORATION
GTCH NASDAQ
101 DYER STREET
PROVIDENCE RI
401/273-7700

GTI CORPORATION
GTI ASE
1060 WILLOW CREEK ROAD
SAN DIEGO CA 92131
619/578-3111

GULTON INDUSTRIES, INC.
GUL NYSE
101 COLLEGE ROAD EAST
PRINCETON NJ 08540
609/452-1811

HABER, INCORPORATED
HABE NASDAQ
470 MAIN ROAD
TOWACO NJ 07082
201/263-0090

HADCO CORPORATION
HDCO NASDAQ
10 MANOR PARKWAY
SALEM NH 03079
603/898-8000

HADRON, INCORPORATED
HDRN NASDAQ
1951 KIDWELL DRIVE
VIENNA VA 22180
703/790-1840

HAEMONETICS CORPORATION
HAEM NASDAQ
400 WOOD ROAD
BRAINTREE MA 02184
617/848-7100

HALE SYSTEMS, INCORPORATED
HSYS NASDAQ
1076 EAST MEADOW CIRCLE
PALO ALTO CA 94303
415/494-6111

HAMILTON DIGITAL CONTROLS, INC.
HDIG NASDAQ
2118 BEECHGROVE PLACE
UTICA NY 13501
315/797-2370

HANDLEMAN COMPANY, INCORPORATED
HDL NYSE
1055 WEST MAPLE ROAD
CLAWSON MI 48017
313/435-3100

HARCO MEDICAL ELECTRONICS, INC.
HMED NASDAQ
17942 SKYPARK CIRCLE STE B
IRVINE CA 92714
714/966-9035

HARCOURT BRACE JOVANOVICH, INC.
HBJ NYSE
757 THIRD AVENUE
NEW YORK NY 10017
212/888-4444

HARNISCHFEGER CORPORATION
HPH NYSE
P.O. BOX 554
MILWAUKEE WI 53201
414/671-4400

HARPER & ROW, PUBLISHERS, INC.
HROW NASDAQ
10 EAST 53RD STREET
NEW YORK NY 10022
212/593-7000

HARRIS & PAULSON, INCORPORATED
HAPI NASDAQ
7887 E. BELLEVIEW, STE 500
ENGLEWOOD CO 80111
303/773-8283

HARRIS CORPORATION
HRS NYSE
P.O. BOX 269
SAN ANTONIO TX 78291
512/344-8000

HARRIS GRAPHICS CORP.
HG NYSE
200 SEMINOLE AVLENUE
MELBOURNE FL 32901
305/676-9400

HAZELTINE CORPORATION
HZ NYSE
500 COMMACK ROAD
COMMACK NY 11725
516/462-5100

HAZLETON LABORATORIES CORP.
HLAB NASDAQ
9200 LEESBURG TURNPIKE
VIENNA VA 22180
703/450-6800

HBO & COMPANY
HBOC NASDAQ
219 PERIMETER CENTER PARKWAY
ATLANTA GA 30346
404/393-6000

HCC INDUSTRIES
HCCI NASDAQ
16311 VENTURA BLVD
ENCINO CA 91436
213/995-4131

HEALTH INFORMATION SYS., INC.
HISI NASDAQ
4522 FORT HAMILTON PARKWAY
BROOKLYN NY 11219
212/435-6300

HEALTH-CHEM CORPORATION
HCH ASE
1107 BROADWAY
NEW YORK NY 10010
212/691-7550

HEALTHDYNE, INCORPORATED
HDYN NASDAQ
2253 N.W. PARKWAY
MARIETTA GA 30067
404/955-9555

HEI, INCORPORATED
HEII NASDAQ
JONATHAN INDUSTRIAL CENTER
CHASKA MN 55318
612/443-2500

HEINICKE INSTRUMENT COMPANY
HEI ASE
3000 TAFT STREET
HOLLYWOOD FL 33021
305/987-6101

HEIZER CORPORATION
HZR ASE
20 N. WACKER DRIVE
CHICAGO IL 60606
312/641-2200

HELIONETICS, INCORPORATED
HILX NASDAQ
17312 EASTMAN STREET
IRVINE CA 92714
714/546-4731

HELIX TECHNOLOGY CORPORATION
HELX NASDAQ
266 SECOND AVENUE
WALTHAM MA 02254
617/890-9292

HEMOTEC, INCORPORATED
HEMO NASDAQ
13 INVERNESS WAY
ENGLEWOOD CO 80112
303/770-1539

HERLEY MICROWAVE SYSTEMS, INC.
HRLY NASDAQ
10 INDUSTRY DRIVE
LANCASTER PA 17603
717/397-2798

HETRA COMPUTER & COMM., INC.
HETC NASDAQ
1151 SOUTH EDDIE ALLEN ROAD
MELBOURNE FL 32901
305/273-7731

HEWLETT-PACKARD COMPANY
HWP NYSE
3000 HANOVER STREET
PALO ALTO CA 94304
415/857-1501

HOGAN SYSTEMS, INCORPORATED
HOGN NASDAQ
5080 SPECTRUM DRIVE
DALLAS TX 75248
214/386-0020

HONEYWELL, INCORPORATED
HON NYSE
HONEYWELL PLAZA
MINNEAPOLIS MN 55408
612/870-5200

HOUGHTON MIFFLIN COMPANY
HTN NYSE
1 BEACON STREET
BOSTON MA 02108
617/725-5000

HUMPHREY, INCORPORATED
HUPH NASDAQ
9212 BALBOA AVENUE
SAN DIEGO CA 92123
714/565-6631

HUNTINGDON RESEARCH CENTER PLC
HRCLY NASDAQ
HUNTINGDON
CAMBS PE186ES
0480 890 431

HURCO MANUFACTURING CO., INC.
HURC NASDAQ
6602 GUION ROAD
INDIANAPOLIS IN 46268
317/293-5309

HYBRIDOMA SCIENCES, INC.
HYBD NASDAQ
11040 CONDOR AVENUE
FOUNTAIN VALLEY CA 92708
714/546-9581

HYBRITECH, INCORPORATED
HYBR NASDAQ
11085 TORREYANA ROAD
SAN DIEGO CA 92121
714/455-6700

HYTEK MICROSYSTEMS, INC.
HTEK NASDAQ
16780 LARK AVENUE
LOS GATOS CA 95030
408/358-1991

IBM CORPORATION
IBM NYSE
ARMONK NY 10504
914/765-1900

ICOT CORPORATION
ICOT NASDAQ
830 MAUDE AVENUE
MOUNTAIN VIEW CA 94043
415/964-4635

ILC TECHNOLOGY, INCORPORATED
ILCT NASDAQ
399 JAVA DRIVE
SUNNYVALE CA 94089
408/745-7900

IMAGIC
IMGC NASDAQ
981 UNIVERSITY AVENUE
LOS GATOS CA 95030
408/399-2200

IMATRON, INCORPORATED
IMAT NASDAQ
389 OYSTER POINT BLVD
S. SAN FRANCISCO CA 94080
415/583-9964

IMEX MEDICAL SYSTEMS, INC.
IMEX NASDAQ
6355 JOYCE DRIVE
GOLDEN CO 80403
303/431-9400

IMM ENERGY SERVICES..., INC.
IMME NASDAQ
3300 S. GESSNER, STE 150
HOUSTON TX 06901
713/266-1177

IMMUNEX CORPORATION
IMNX NASDAQ
51 UNIVERSITY STREET, STE 600
SEATTLE WA 98101
206/587-0430

IMMUNOGENETICS (IMMUGEN), INC.
IGEN NASDAQ
2285 E. LANDIS AVENUE
VINELAND NJ 08360
609/691-2411

IMMUNUNO NUCLEAR CORPORATION
INUC NASDAQ
1951 NORTHWESTERN AVENUE
STILLWATER MN 55082
612/439-9710

IMREG INCORPORATED
IMRGA NASDAQ
144 ELK PLACE #1400
NEW ORLEANS LA 70112
504/523-2875

INACOMP COMPUTER CENTERS
INAC NASDAQ
1824 WEST MAPLE ROAD
TROY MI 48084
313/649-0910

INDUSTRIAL SOLID STATE CONTROLS
ISSC NASDAQ
P.O. BOX 934
YORK PA 17405
717/848-1151

INFORMATICS GENERAL CORP.
IG NYSE
21031 VENTURA BLVD
WOODLAND HILLS CA 91364
213/887-9040

INFORMATION SCIENCE, INC.
INSI NASDAQ
95 CHESTNUT RIDGE ROAD
MONTVALE NJ 07645
201/391-1600

INFORMATION INTERNATIONAL, INC.
IINT NASDAQ
5933 SLAUSON AVENUE
CULVER CITY CA 90230
213/390-8611

INFORMATION DISPLAYS, INC.
IDPY NASDAQ
28 KAYSAL COURT
ARMONK NY 10504
914/273-5755

INFOTRON SYSTEMS CORPORATION
INFN NASDAQ
CHERRY HILL INDUSTRIAL CENTER
CHERRY HILL NJ 08003
609/424-9400

INFRARED INDUSTRIES, INC.
INFR NASDAQ
P.O. BOX 989
SANTA BARBARA CA 93102
805/684-4181

INNOVATIVE SOFTWARE, INC.
INSO NASDAQ
9300 WEST 110TH STREET #380
OVERLAND PARK KS 66212
913/383-1089

INSILCO CORPORATION
INR NYSE
1000 RESEARCH PKWY
MERIDEN CT 06450
203/634-2000

INSTACOM, INCORPORATED
ICOM NASDAQ
7610 STEMMONS FREEWAY
DALLAS TX 75247
214/631-1505

INSTRON CORPORATION
ISN ASE
100 ROYALL STREET
CANTON MA 02021
617/828-2500

INTECOM, INCORPORATED
INCM NASDAQ
601 INTECOM DRIVE
ALLEN TX 75002
214/727-9141

INTEGRATED SOFTWARE SYS. CORP (ISSCO)
ISCX NASDAQ
10505 SORRENTO VALLEY ROAD
SAN DIECO CA 92121
619/452-0170

INTEGRATED DEVICE TECHNOLOGY
IDII NASDAQ
3236 SCOTT BLVD
SANTA CLARA CA 95051
408/727-6116

INTEGRATED AUTOMATION, INC.
INAU NASDAQ
2121 ALLSTON WAY
BERKELEY CA 94704
415/843-8227

INTEGRATED GENETICS, INC.
INGN NASDAQ
51 NEW YORK AVENUE
FRAMINGHAM MA 01701
617/875-1336

INTEL CORPORATION
INTC NASDAQ
3065 BOWERS AVENUE
SANTA CLARA CA 95051
408/987-8080

INTELLICORP
INAI NASDAQ
707 LAUREL STREET
MENLO PARK CA 94025
415/323-8300

INTELLIGENT COMMUNICATIONS NETWORKS, INC.
ICNT NASDAQ
2772 JOHNSON DRIVE
VENTURA CA 93003
805/654-1616

INTELLIGENT SYSTEMS CORPORATION
INTS NASDAQ
225 TECHNOLOGY PARK
ATLANTA/NORCROSS GA 30092
404/449-5961

INTER'L GAME TECHNOLOGY
IGAM NASDAQ
520 SOUTH ROCK BLVD.
RENO NV 89502
702/323-5060

INTER'L GENETIC ENGINEERING
IGEI NASDAQ
1701 COLORADO AVENUE
SANTA MONICA CA 90404
213/829-7681

INTER'L MICROELECTRONIC PROD'S
IMPX NASDAQ
2830 NORTH FIRST STREET
SAN JOSE CA 95134
408/262-9100

INTER'L MOBILE MACHINES CORP.
IMMC NASDAQ
100 N. 20TH STREET
PHILADELPHIA PA 19103
215/569-3880

INTER'L REMOTE IMAGING SYSTEMS
IRIS NASDAQ
9232 DEERING AVENUE
CHATSWORTH CA 91311
213/709-1244

INTER'L ROBOMATION/INTELLIGENCE
ROBTC NASDAQ
2281 LAS PALMAS DRIVE
CARLSBAD CA 92008
619/438-4424

INTER'L TECHNOLOGY CORP.
ITCP NASDAQ
336 W ANAHEIM STREET
WILMINGTON CA 90744
213/830-1781

INTER-TEL, INCORPORATED
INTL NASDAQ
3232 W. VIRGINIA AVENUE
PHOENIX AZ 85009
602/269-5091

INTERACTION SYSTEMS, INC.
ISIIA NASDAQ
24 MUNROE STREET
NEWTONVILLE MA 02160
617/964-5300

INTERACTIVE RADIATION (INRAD)
INRD NASDAQ
181 LEGRAND AVENUE
NORTHVALE NJ 07647
201/767-1910

INTERACTIVE SYSTEMS CORPORATION
ISCO NASDAQ
5500 SOUTH SYCAMORE ST
LITTLETON CO 80120
303/797-2400

INTERFACE SYSTEMS, INCORPORATED
INTF NASDAQ
5855 INTERFACE DRIVE
ANN ARBOR MI 48103
313/769-5900

INTERFERON SCIENCES, INC.
IFSC NASDAQ
783 JERSEY AVENUE
NEW BRUNSWICK NJ 08901
201/249-3232

INTERGATED CIRCUITS, INC.
ICTM NASDAQ
13256 NORTHUP WAY
BELLEVUE WA 98005
206/747-8556

INTERGRAPH CORPORATION
INGR NASDAQ
ONE MADISON INDUSTRIAL PARK
HUNTSVILLE AL 35807
205/772-2000

INTERMAGNETICS GENERAL CORP.
INMA NASDAQ
CHARLES IND PARK/NEW KARNER RD
GUILDERLAND NY 12084
518/456-5456

INTERMEC CORPORATION
INTR NASDAQ
4405 RUSSEL ROAD
LYNWOOD WA
206/743-7036

INTERMEDICS, INCORPORATED
ITM NYSE
P.O. BOX 617
FREEPORT TX 77541
713/233-8611

INTERMETRICS, INCORPORATED
IMET NASDAQ
733 CONCORD AVENUE
CAMBRIDGE MA 02138
617/661-1840

INTERNATIONAL CLINICAL LABORATORIES, INC.
ICLB NASDAQ
P.O. BOX 24027
NASHVILLE TN 37202
615/327-1025

INTERTEC DATA SYSTEMS CORP.
IDC ASE
2300 BROAD RIVER ROAD
COLUMBIA SC 29210
803/798-9100

INVENTION DESIGN ENGINEERING
IDEA NASDAQ
500 ALASKA STREET
TORRANCE CA 90503
213/320-9462

IOMEGA CORPORATION
IOMG NASDAQ
4646 SOUTH 1500 WEST
OGDEN UT 84403
801/392-7581

IPL SYSTEMS, INCORPORATED
IPLSA NASDAQ
1370 MAIN STREET
WALTHAM MA 02254
617/890-6620

IRT CORPORATION
IX ASE
7650 CONVOY COURT
SAN DIEGO CA 92111
714/565-7171

IRVINE SENSORS CORPORATION
IRSN NASDAQ
3001 REDHILL AVENUE
COSTA MESA CA 92626
714/549-8211

ISC SYSTEMS CORPORATION
ISCS NASDAQ
9922 MONTGOMERY
SPOKANE WA 99288
509/536-5050

ISOMEDIX, INCORPORATED
ISMX NASDAQ
11 APOLLO DRIVE
WHIPPANY NJ 07981
201/887-4700

ISOMET CORPORATION
IOMT NASDAQ
5263 PORT ROYAL ROAD
SPRINGFIELD VA 22151
703/321-8301

ITT CORPORATION
ITT NYSE
320 PARK AVENUE
NEW YORK NY 10022
212/752-6000

IU INTERNATIONAL CORPORATION
IU NYSE
1105 N. MARKET STREET
WILMINGTON DE 19801
302/571-5000

JEC LASERS, INCORPORATED
JECL NASDAQ
59 NORTH 5TH STREET
SADDLE BROOK NJ 07662
201/843-6600

JOHNSON CONTROLS, INC.
JCI NYSE
P.O. BOX 591
MILWAUKEE WI 53201
414/228-3155

JOY MFG COMPANY
JOY NYSE
1200 HENRY W. OLVIER BLDG.
PITTSBURGH PA 15222
412/562-4500

K-TRON INTERNATIONAL, INC.
KTII NASDAQ
7975 N. HAYDEN RD, STE D-360
SCOTTSDALE AZ 85258
602/998-0900

KAMAN CORPORATION
KAMNA NASDAQ
BLUE HILLS AVENUE
BLOOMFIELD CT 06002
203/243-8311

KAYPRO CORPORATION
KAYP NASDAQ
533 STEVENS AVENUE
SOLANA BEACH CA 92075
619/755-1134

KCR TECHNOLOGY, INCORPORATED
KCRT NASDAQ
100 PRESTIGE PARK ROAD
EAST HARTFORD CT 06108
203/289-8618

KEANE, INCORPORATED
KEAN NASDAQ
210 COMMERCIAL STREET
BOSTON MA 02109
617/742-5210

KEARNEY-NATIONAL, INCORPORATED
KERN NASDAQ
200 PARK AVENUE
NEW YORK NY 10166
212/972-9590

KEVEX CORPORATION
KEVX NASDAQ
1101 CHESS DRIVE
FOSTER CITY CA 94404
415/573-5866

KEVLIN MICROWAVE CORPORATION
KVLM NASDAQ
26 CONN STREET
WOBURN MA 01801
617/935-4800

KEWAUNEE SCI. EQUIPMENT CORP.
KEQU NASDAQ
STATESVILLE NC 28677
704/873-7202

KEY IMAGE SYSTEMS, INC.
KIMGA NASDAQ
20100 PLUMMER STREET
CHATSWORTH CA 91311
213/993-1911

KEY TRONIC CORPORATION
KTCC NASDAQ
SPOKANE INDUSTRIAL PARK, BLD 1
SPOKANE WA 99216
509/928-8000

KLA INSTRUMENTS CORPORATION
KLAC NASDAQ
2051 MISSION COLLEGE BLVD
SANTA CLARA CA 95054
408/988-6100

KMS INDUSTRIES, INCORPORATED
KMSI NASDAQ
3941 RESEARCH PARK DRIVE
ANN ARBOR MI 48108
313/769-1100

KNOGO CORPORATION
KNO ASE
100 TECH STREET
HICKSVILLE NY 11801
516/822-4200

KOLFF MEDICAL, INCORPORATED
KOLF NASDAQ
825 NORTH 300 WEST
SALT LAKE CITY UT 84103
801/531-7022

KOLLMORGEN CORPORATION
KOL NYSE
66 GATE HOUSE ROAD
STAMFORD CT 06902
203/327-7222

KRATOS, INCORPORATED
KTOS NASDAQ
3333 N. TORREY PINES COURT
LA JOLLA CA 92037
619/455-9020

KULICKE AND SOFFA IND., INC.
KLIC NASDAQ
507 PRUDENTIAL ROAD
HORSHAM PA 19044
215/674-2800

L/F TECHNOLOGIES, INCORPORATED
LFTI NASDAQ
2800 LOCKHEED WAY
CARSON CITY NV 89701
702/883-7611

LAIDLAW INDUSTRIES, INC.
LWSI NASDAQ
15 SPINNING WHEEL ROAD
HINSDALE IL 60521
312/877-8181

LAM RESEARCH CORPORATION
LRCX NASDAQ
47531 WARM SPRINGS BLVD
FREMONT CA 94539
415/659-0200

LANIER BUSINESS PRODUCTS
LBP NYSE
1700 CHANTILLY DRIVE, N.E.
ATLANTA GA 30324
404/329-8000

LASER CORPORATION
LSER NASDAQ
1832 S. 3850 WEST
SALT LAKE CITY UT 84104
801/972-1311

LASER PHOTONICS, INCORPORATED
LAZR NASDAQ
2025 PALMRIDGE WAY
ORLANDO FL 32809
305/851-7424

LASER PRECISION CORPORATION
LASR NASDAQ
17819 GILLETTE AVENUE
IRVINE CA 92714
714/660-8801

LASERMED CORPORATION
LAMD NASDAQ
151 KALMUS, STE H3
COSTA MESA CA 92626
714/432-9660

LASERS FOR MEDICINE, INCORPORATED
LFMI NASDAQ
77 ARKAY DRIVE
HAUPPAUGE NY 11788
516/231-7727

LASERTECHNICS, INCORPORATED
LASX NASDAQ
6007 OSUNA RD N.E.
ALBUQUERQUE NM 87109
503/883-5353

LEAR SIEGLER, INCORPORATED
LSI NYSE
2850 OCEAN PARK BLVD
SANTA MONICA CA 90406
213/452-6000

LEE DATA CORPORATION
LEDA NASDAQ
7075 FLYING CLOUD DRIVE
EDEN PRAIRIE MN 55344
612/828-0300

LEVIN INTERNATIONAL CORP.
LEVN NASDAQ
224 EAST 49TH ST.
NEW YORK NY 10017
212/935-9620

LEXICON CORPORATION
LEXI NASDAQ
1541 NW 65TH AVENUE
FT. LAUDERDALE FL 33313
305/792-4400

LEXIDATA CORPORATION
LEXD NASDAQ
755 MIDDLESEX TURNPIKE
BILLERICA MA 01865
617/663-8550

LFE CORPORATION
LFE NYSE
55 GREEN STREET
CLINTON MA 01510
617/365-4511

LIFE SCIENCES, INCORPORATED
LFSC NASDAQ
2900 72ND STREET NORTH
ST PETERSBURG FL 33710
813/345-9371

LIFE TECHNOLOGIES, INCORPORATED
LTEK NASDAQ
CHAGRIN FALLS OH 44022
216/247-4300

LIFELINE SYSTEMS, INCORPORATED
LIFL NASDAQ
400 MAIN STREET
WALTHAM MA 02254
617/893-2211

LINEAR CORPORATION
LINE NASDAQ
347 SOUTH GLASGOW AVENUE
INGLEWOOD CA 90301
213/649-0222

LITTON INDUSTRIES, INCORPORATED
LIT NYSE
360 N. CRESCENT DRIVE
BEVERLY HILLS CA 90210
213/273-7860

LOCKHEED CORPORATION
LK NYSE
2555 NORTH HOLLYWOOD WAY
BURBANK CA 91520
213/847-6121

LOGETRONICS, INCORPORATED
LOGE NASDAQ
7001 LOISDALE ROAD
SPRINGFIELD VA 22150
703/971-1400

LOGICON, INCORPORATED
LGN ASE
3701 SKYPARK DRIVE
TORRANCE CA 90505
213/373-0220

LOGOS SCIENTIFIC, INCORPORATED
LOGS NASDAQ
700 WEST SUNSET ROAD
HENDERSON NV 89015
702/565-1383

LORAL CORPORATION
LOR NYSE
600 THIRD AVENUE
NEW YORK NY 10016
212/697-1105

LOTUS DEVELOPMENT CORPORATION
LOTS NASDAQ
55 WHEELER STREET
CAMBRIDGE MA 02138
617/492-7171

LSI LOGIC CORPORATION
LLSI NASDAQ
1601 MC CARTHY BLVD
MILPITAS CA 95035
408/263-9494

LTX CORPORATION
LTXX NASDAQ
145 UNIVERSITY AVENUE
WESTWOOD MA 02090
617/329-7550

LUNDY ELECTRONICS & SYS., INC.
LDY ASE
GLEN HEAD NY 11545
516/671-9000

LYON METAL PRODUCTS, INC.
LYON NASDAQ
P.O. BOX 671
AURORA IL 60507
312/892-8941

M/A-COM, INCORPORATED
MAI NYSE
SOUTH AVENUE
BURLINGTON MA 01803
617/272-9600

MACHINE TECHNOLOGY, INC.
MTEC NASDAQ
20 LESLIE COURT
WHIPPPANY NJ 07981
201/386-0600

MACMILLAN, INCORPORATED
MLL NYSE
866 THIRD AVENUE
NEW YORK NY 10022
212/935-2000

MACNEAL-SCHWENDLER CORP. (THE)
MNS AMEX
815 COLORADO BLVD.
LOS ANGELES CA 90041
213/258-9100

MAGNETIC CONTROLS COMPANY
MGNE NASDAQ
4900 WEST 78TH STREET
MINNEAPOLIS MN 55435
612/835-6800

MAGNETIC TECHNOLOGIES CORP.
MTCC NASDAQ
60 SAGINAW DRIVE
ROCHESTER NY 14623
716/244-1343

MANAGEMENT ASSISTANCE, INC.
M NYSE
560 LEXINGTON AVENUE
NEW YORK NY 10022
212/909-1400

MANAGEMENT SCIENCE AMERICA, INC
MSAI NASDAQ
3445 PEACHRTREE ROAD, NE
ATLANTA GA 30326
404/239-2000

MARK CONTROLS CORPORATION
MK NYSE
1900 DEMPSTER STREET
EVANSTON IL 60204
312/866-8840

MARQUEST MEDICAL PRODUCTS, INC.
MMPI NASDAQ
112 INVERNESS CIRCLE STE A
ENGLEWOOD CO 80112
303/770-4835

MARSHALL INDUSTRIES
MI ASE
9674 TELSTAR AVENUE
EL MONTE CA 91731
213/442-7204

MARTIN MARIETTA CORPORATION
ML NYSE
6801 ROCKLEDGE DRIVE
BETHESDA MD 20817
301/897-6000

MASSCOMP
MSCP NASDAQ
ONE TECHNOLOGY PK
WESTFORD MA 01886
617/692-6200

MASSTOR SYSTEMS CORPORATION
MSCO NASDAQ
541 LAKESIDE DRIVE
SUNNYVALE CA 94086
408/737-2500

MATERIALS RESEARCH CORPORATION
MTL ASE
RT 303
ORANGEBURY NY 10962
914/359-4200

MATHEMATICAL APPLICATIONS, INC.
MAGC NASDAQ
3 WESTCHESTER PLAZA
ELMSFORD NY 10523
914/592-4646

MATRIX CORPORATION
MAX ASE
ONE RAMLAND ROAD
ORANGEBURG NY 10962
914/365-0190

MATRIX SCIENCE CORPORATION
MTRX NASDAQ
455 MAPLE AVENUE
TORRANCE CA 90503
213/328-0271

MATTEL, INCORPORATED
MAT NYSE
5150 ROSECRANS AVENUE
HAWTHORNE CA 90250
213/978-5150

MAXCO, INCORPORATED
MAXC NASDAQ
P.O. BOX 26127
LANSING MI 48909
517/393-7423

MAXWELL LABORATORIES, INC.
MXWL NASDAQ
8835 BALBOA AVENUE
SAN DIEGO CA 92123
619/279-5100

MCDONNELL DOUGLAS CORPORATION
MD NYSE
P.O. BOX 516
ST. LOUIS MO 63166
314/232-0232

MCGRAW-HILL, INCORPORATED
MHP NYSE
1221 AVENUE OF THE AMERICAS
NEW YORK NY 10020
212/512-2000

MCI COMMUNICATIONS CORPORATION
MCIC NASDAQ
1133 19TH ST., N.W.
WASHINGTON DC 20036
202/872-1600

MEASUREX CORPORATION
MX NYSE
ONE RESULTS WAY
CUPERTINO CA 95014
408/255-1500

MEDICAL DYNAMICS, INCORPORATED
MEDY NASDAQ
14 INVERNESS DRIVE, EAST
ENGLEWOOD CO 80110
303/770-2990

MEDICAL ELECTRONICS CORP. OF AM
MECA NASDAQ
6595 O'DELL PLACE
BOULDER CO 80301
303/530-3845

MEDICAL GRAPHICS CORPORATION
MGCC NASDAQ
501 WEST COUNTY RD E
ST PAUL MN 55112
612/484-4874

MEDIFLEX SYSTEMS, INCORPORATED
MFLX NASDAQ
990 GROVE STREET
EVANSTON IL 60201
312/866-1500

MEDTRONIC, INCORPORATED
MDT NYSE
3055 OLD HIGHWAY EIGHT
MINNEAPOLIS MN 55440
612/574-4000

MEGADATA CORPORATION
MDTA NASDAQ
35 ORVILLE DRIVE
BOHEMIA NY 11716
516/589-6800

MENTOR CORPORATION
MNTR NASDAQ
1499 W. RIVER RD. N.
MINNEAPOLIS MN 55411
612/588-4685

MENTOR GRAPHICS CORPORATION
MENT NASDAQ
8500 SW CREEKSIDE PLACE
BEAVERTON OR 97005
503/626-7000

METHODE ELECTRONICS, INC.
METHB NASDAQ
7444 WEST WILSON AVENUE
HARWOOD HEIGHTS IL 60656
312/867-9600

METROMEDIA, INCORPORATED
MET NYSE
1 HARMON PLAZA
SECAUCUS NJ 07094
201/348-3244

MICOM SYSTEMS, INCORPORATED
MICS NASDAQ
9551 PRONDALE AVENUE
CHATSWORTH CA 91311
213/998-8844

MICRO D, INCORPORATED
MCRD NASDAQ
17406 MT. CLIFFWOOD CIRCLE
FOUNTAIN VALLEY CA 92708
714/540-4781

MICRO MASK, INCORPORATED
MCRO NASDAQ
695 VAGUEROS AVENUE
SUNNYVALE CA 94086
408/245-7342

MICRO Z CORPORATION
MICZ NASDAQ
11754 WILSHIRE BLVD.
LOS ANGELES CA 90025
213/478-6093

MICROBIOLOGICAL SCIENCES, INC.
MBLS NASDAQ
771 MAIN STREET
WEST WARWICK RI 02893
401/828-5250

MICROCOMPUTER MEMORIES, INC.
MCMI NASDAQ
7444 VALJEAN AVENUE
VAN NUYS CA 91406
818/782-2222

MICRODYNE CORPORATION
MCDY NASDAQ
P.O. BOX 7213
OCALA FL 32672
904/687-4633

MICROFAST SOFTWARE CORP.
FAST NASDAQ
420 L STREET, SUITE 407
ANCHORAGE AK 99501
907/277-2211

MICROMARKETING INTER'L, INC.
MMAR NASDAQ
477 EAST 3RD STREETS
WILLIAMSPORT PA 17701
717/327-9575

MICRON TECHNOLOGY, INC.
DRAM NASDAQ
2805 E. COLUMBIA RD.
BOISE ID 83706
208/383-4000

MICROS SYSTEMS, INCORPORATED
MCRS NASDAQ
6901-B DISTRIBUTION DRIVE
BELTSVILLE MD 20705
301/937-9080

MICROSEMI CORPORATION
MSCC NASDAQ
2830 SOUTH FAIRVIEW STREET
SANTA ANA CA 92704
714/979-8220

MIDLAND CAPITAL CORPORATION
MCAP NASDAQ
950 3RD AVENUE
NEW YORK NY 10022
212/577-0750

MIDLAND-ROSS CORPORATION
MLR NYSE
20600 CHAGRIN BLVD
CLEVELAND OH 44122
216/491-8400

MILLER TECH. & COMM. CORP.
MECC NASDAQ
4837 E. MCDOWELL ROAD
PHOENIX AZ 85008
602/254-1129

MILLICOM, INCORPORATED
MILL NASDAQ
153 EAST 53 STREET, STE 5500
NEW YORK NY 10022
212/355-3574

MILLIPORE CORPORATION
MILI NASDAQ
80 ASHBY ROAD
BEDFORD MA 01730
617/275-9200

MILTON BRADLEY COMPANY
MB NYSE
111 MAPLE STREET
SPRINGFIELD MA 01105
413/525-6411

MILTON ROY COMPANY
MRC NYSE
ONE PLAZA PLACE NE
ST PETERSBURG FL 33701
813/823-4444

MINCOMP CORPORATION
MCOM NASDAQ
5680 S. SYRACUSE CIRCLE
ENGLEWOOD CO 80111
303/779-3063

MINISCRIBE CORPORATION
MINY NASDAQ
1871 LEFTHAND CIRCLE
LONGMONT CO 80501
303/651-6000

MITRAL MEDICAL INTER'L, INC.
MMI NASDAQ
4050 YOUNGFIELD STREET
WHEAT RIDGE CO 80033
303/431-6000

MMI MEDICAL, INCORPORATED
MMIM NASDAQ
1902 ROYALTY DR., STE 220
POMONA CA 91767
714/620-0391

MOBILE COMM. CORP. OF AMERICA
MCCAB NASDAQ
CAPITAL TOWERS BLDG, STE 1500
JACKSON MS 39281
601/969-1200

MOBOT CORPORATION
MBOT NASDAQ
980 BUENOS AVENUE
SAN DIEGO CA 92110
714/275-4300

MODULAR COMPUTER SYSTEMS, INC.
MSY NYSE
1650 W. MCNAB ROAD
FT. LAUDERDALE FL 33309
305/974-1380

MOHAWK DATA SCIENCES CORP.
MDS NYSE
7 CENTURY DRIVE
PARSIPPANY NJ 07054
201/540-9080

MOLECULAR GENETICS, INC.
MOGN NASDAQ
10320 BREN ROAD EAST
MINNETONKA MN 55343
612/935-7335

MOLEX, INCORPORATED
MOLX NASDAQ
2222 WELLINGTON COURT
LISLE IL 60532
312/969-4550

MONCHIK-WEBER CORPORATION (THE)
MWCH NASDAQ
11 BROADWAY
NEW YORK NY 10004
212/269-5460

MONITOR LABS, INCORPORATED
MLAB NASDAQ
10180 SCRIPPS RANCH BLVD.
SAN DIEGO CA 92131
714/578-5060

MONOCLONAL ANTIBODIES, INC.
MABS NASDAQ
2319 CHARLESTON ROAD
MOUNTAIN VIEW CA 94043
415/960-1320

MONOLITHIC MEMORIES, INC.
MMIC NASDAQ
1165 E. ARQUES STREET
SUNNYVALE CA 94086
408/970-9700

MONOSIL, INCORPORATED
MSIL NASDAQ
3060 RAYMOND STREET
SANTA CLARA CA 95050
408/727-6562

MONSANTO COMPANY
MTC NYSE
800 N. LINDBERGH BLVD
ST LOUIS MO 63166
314/694-1000

MOORE PRODUCTS COMPANY
MORP NASDAQ
SPRING HOUSE PA 19477
215/646-7400

MOTOROLA, INCORPORATED
MOT NYSE
1303 E. ALGONQUIN RD
SCHAUMBURG IL 60196
312/397-5000

MPSI GROUP, INC. (THE)
MPSG NASDAQ
8282 S. MEMORIAL DRIVE
TULSA OK 74133
918/250-9611

MSI DATA CORPORATION
MSI ASE
340 FISCHER AVENUE
COSTA MESA CA 92626
714/549-6000

MTS SYSTEMS CORPORATION
MTSC NASDAQ
8055 MITCHELL ROAD
EDEN PRAIRIE MN 55344
612/937-4000

MULTI SOLUTIONS, INCORPORATED
MULT NASDAQ
660 WHITEHEAD ROAD
LAWRENCEVILLE NJ 08648
609/695-1337

NANOMETRICS, INCORPORATED
NANO NADSAQ
930 WEST MAUDE AVENUE
SUNNYVALE CA 94086
408/735-1044

NAPCO SECURITY SYSTEMS, INC.
NSSC NASDAQ
6 DITOMAS COURT
COPIAGUE NY 11726
800/645-9445

NARCO SCIENTIFIC, INCORPORATED
NAO NYSE
455 PENNSYLVANIA AVE
FORT WASHINGTON PA 19034
215/641-5800

NARDA MICROWAVE CORP. (THE)
NRD ASE
435 MORELAND ROAD
HAUPPAUGE NY 11788
516/231-1700

NARRAGANSETT CAPTIAL CORP.
NARR NASDAQ
40 WESTMINISTER STREET
PROVIDENCE RI 02903
401/751-1000

NATIONAL COMPUTER SYSTEMS
NLCS NASDAQ
4401 W. 76TH STREET
EDINA MN 55435
612/830-7600

NATIONAL DATA CORPORATION
NDTA NASDAQ
ONE NATIONAL DATA PLAZA
ATLANTA GA 30329
404/329-8500

NATIONAL DATA COMMUNICATIONS
NDAC NASDAQ
5440 HARVEST HILL ROAD
DALLAS TX 75230
214/386-0600

NATIONAL EDUCATION CORP.
NEC NYSE
4361 BIRCH STREET
NEWPORT BEACH CA 92660
714/546-7360

NATIONAL ENVIRONMENTAL CONTROLS
NECT NASDAQ
912 DAVID DRIVE
METAIRIE LA 70003
504/733-0586

NATIONAL MICRONETICS, INC.
NMIC NASDAQ
5600 KEARNY MESA ROAD
SAN DIEGO CA 92111
714/279-7500

NATIONAL SEMICONDUCTOR CORP.
NSM NYSE
2900 SEMICONDUCTOR DRIVE
SANTA CLARA CA 95051
408/721-5000

NATIONAL TECHNICAL SYSTEMS
NTSC NASDAQ
15720 VENTURA BLVD
ENCINO CA 91436
213/789-7793

NATIONAL VIDEO CENTERS, INC.
NVID NASDAQ
105 ORVILLE DRIVE
BOHEMIA NY 11716
516/567-9500

NAUTILUS FUND
NTLS NASDAQ
24 FEDERAL STREET
BOSTON MA 02110
617/482-8260

NBI, INCORPORATED
NBI NYSE
1695 38TH STREET
BOULDER CO 80301
303/444-5710

NCA CORPORATION
NCAC NASDAQ
388 OAKMEAD PARKWAY
SUNNYVALE CA 94086
408/245-7990

NCR CORPORATION
NCR NYSE
1700 S PATTERSON BLVD
DAYTON OH 45479
513/445-5000

NEO-BIONICS, INCORPORATED
NEOB NASDAQ
4117 MONTGOMERY N.E.
ALBUQUERQUE NM 87109
505/883-7204

NETWORD, INCORPORATED
NTWD NASDAQ
6801 KENILWORTH AVENUE
RIVERDALE MD 20737
301/699-0100

NETWORK SYSTEMS CORPORATION
NSCO NASDAQ
7600 BOONE AVE, NORTH
BROOKLYN PARK MN 55428
612/425-2202

NETWORKS ELECTRONIC CORP.
NWRK NASDAQ
9750 DESOTO AVENUE
CHATSWORTH CA 91311
213/341-0440

NEW BRUNSWICK SCIENTIFIC CO.
NBSC NASDAQ
44 TALMADGE ROAD
EDISON NJ 08818
201/287-1200

NEW WORLD COMPUTER CO., INC.
NEWW NASDAQ
2805 MC GRAW AVENUE
IRVINE CA 92714
714/566-9320

NEWPORT CORPORATION
NEWP NASDAQ
18235 MT. BALDY CIRCLE
FOUNTAIN VALLEY CA 92708
714/963-9811

NEWPORT ELECTRONICS, INC.
NEWE NASDAQ
630 EAST YOUNG STREET
SANTA ANA CA 92705
714/540-4914

NICOLET INSTRUMENT CORPORATION
NIC NYSE
P.O. BOX 4451
MADISON WI 53791
608/721-3333

NMR OF AMERICA, INCORPORATED
NMRR NASDAQ
95 MADISON AVENUE, #100
MORRISTOWN NJ 07960
201/539-1082

NORDSON CORPORATION
NDSN NASDAQ
P.O. BOX 151
AMHERST OH 44001
216/988-9411

NORTH AMERICAN BIOLOGICALS INC.
NBIO NASDAQ
16500 NORTHWEST 15 AVE
MIAMI FL 33169
305/625-5303

NORTH AMERICAN PHILIPS CORP.
NPH NYSE
100 E. 42ND STREET
NEW YORK NY 10017
212/697-3600

NORTHERN DATA SYSTEMS, INC.
NDSI NASDAQ
48 CONSTITUTION DRIVE
BEDFORD NH 03102
603/472-8855

NORTHROP CORPORATION
NOC NYSE
1800 CENTURY PARK EAST
LOS ANGELES CA 90067
213/553-6262

NOVAMETRIX MEDICAL SYSTEMS
NMTX NASDAQ
ONE BARNES INDUSTRIAL PARK
WALLINGFORD CT 06492
203/265-7701

NOVAN ENERGY, INCORPORATED
NOVN NASDAQ
1630 NORTH 63RD STREET
BOULDER CO 80301
303/447-9193

NOVAR ELECTRONICS CORPORATION
NOVR NASDAQ
24 BROWN STREET
BARBERTON OH 44203
216/745-0074

NU-MED, INCORPORATED
NUMS NASDAQ
16633 VENTURA BLVD.
ENCINO CA 91436
213/990-2000

NUCLEAR DATA, INCORPORATED
NDI ASE
HAMILTON LAKES 500 PARK BLVD
ITASCA IL 60143
312/773-0200

NUCLEAR METALS, INCORPORATED
NUCM NASDAQ
2229 MAIN STREET
CONCORD MA 01742
617/369-5410

NUCLEAR PHARMACY, INC.
NURX NASDAQ
4272 BALLOON PARK RD. N.E.
ALBUQUERQUE NM 87109
505/345-3551

NUCLEAR SUPPORT SERVICES
NSSI NASDAQ
208 BRIARCREST SQUARE
HERSHEY PA 17033
717/533-4835

NUCLEAR SYSTEMS, INCORPORATED*
NUSYQ NASDAQ
924 JOPLIN ST
BATON ROUGE LA 70802
504/388-0801

NUMERAX, INCORPORATED
NMRX NASDAQ
230 W. PASSAIC STREET
MAYWOOD NJ 07607
201/368-0170

O.C.G. TECHNOLOGY, INCORPORATED
OCGT NASDAQ
16 WEST 61ST STREET
NEW YORK NY 10023
212/541-5470

OAK INDUSTRIES, INCORPORATED
OAK NYSE
16935 W. BERNARDO DRIVE
RANCHO BERNARDO CA 92127
619/485-9300

OBJECT RECOGNITION SYS., INC.
ORSI NASDAQ
521 FIFTH AVENUE, 17TH FLOOR
NEW YORK NY 10175
212/581-4200

OCEAN TECHNOLOGY, INCORPORATED
OTII NASDAQ
2835 NORTH NAOMI STREET
BURBANK CA 91504
213/849-7111

ODETICS, INCORPORATED
ODEX NASDAQ
1380 SOUTH ANAHEIM BLVD.
ANAHEIM CA 92805
714/774-5000

OEA, INCORPORATED
OEA ASE
P.O. BOX 10488
DENVER CO 80210
303/693-1248

OMNIMEDICAL
OMNIQ NASDAQ
355 WOODHEAD DRIVE
NORTHBROOK IL 60062
312/564-5510

OMNITRONICS RESEARCH CORP.
ORCS NASDAQ
2775 BARBER ROAD
NORTON OH 44203
216/848-4192

ON-LINE SOFTWARE INTER'L, INC.
OSII NASDAQ
2 EXECUTIVE PARK
FORT LEE NJ 07024
201/592-0009

ONYX + IMI, INCORPORATED
ONIX NASDAQ
25 E. TRUMBLE ROAD
SAN JOSE CA 95131
408/946-6330

OPTELECOM, INCORPORATED
OPTC NASDAQ
15940 LUANNE DRIVE
GAITHERSBURG MD 20877
301/840-2121

OPTICAL COATING LAB., INC.
OCLI NASDAQ
P.O. BOX 1599
SANTA ROSA CA 95402
707/545-6440

OPTICAL RADIATION CORPORATION
ORCO NASDAQ
1300 OPTICAL DRIVE
AZUSA CA 91702
818/969-3344

OPTICAL SPECIALTIES, INC.
OSIX NASDAQ
4281 TECHNOLOGY DRIVE
FREMONT CA 94538
415/490-6400

ORFA CORP. OF AMERICA
ORFA NASDAQ
800 KINGS HIGHWAY, STE 216
CARYHILL NJ 08034
609/482-2300

ORION RESEARCH, INC.
ORIR NASDAQ
840 MEMORIAL DRIVE
CAMBRIDGE MA 02139
617/864-5400

OSMONICS, INCORPORATED
OSMO NASDAQ
5951 CLEARWATER DRIVE
MINNETONKA MN 55343
612/933-2277

PACIFIC SCIENTIFIC COMPANY
PSX NYSE
1350 S. STATE COLLEGE
ANAHEIM CA 92803
714/535-8141

PALL CORPORATION
PLL ASE
30 SEA CLIFF AVENUE
GLEN COVE, L.I. NY 11542
516/671-4000

PANSOPHIC SYSTEMS, INCORPORATED
PANS NASDAQ
709 ENTERPRISE DRIVE
OAK BROOK IL 60521
312/986-6000

PAR TECHNOLOGY CORPORATION
PARR NASDAQ
SENECA PLAZA, ROUTE 5
NEW HARTFORD NY 13413
315/738-0600

PARADYNE CORPORATION
PDN NYSE
8550 ULMERTON ROAD
LARGO FL 33541
813/530-2000

PARAHO DEVELOPMENT CORPORATION
PAHO NASDAQ
183 INVERNESS DRIVE W #300A
DENVER CO 80112
303/694-4949

PARK ELECTROCHEMICAL CORP.
PKE NYSE
475 NORTHERN BLVD
GREAT NECK NY 11021
516/466-5700

PARLEX CORPORATION
PRLX NASDAQ
145 MILK STREET
METHUEN MA 01844
617/685-4341

PATIENT TECHNOLOGY, INC.
PTIX NASDAQ
80 DAVID'S DRIVE
HAUPPAUGE NY 11788
516/435-0905

PAYCHEX, INCORPORATED
PAYX NASDAQ
275 LAKE AVENUE
ROCHESTER NY 14608
716/647-3510

PDA ENGINEERING
PDAS NASDAQ
1560 BROOKHOLLOW DR.
SANTA ANA CA 92705
714/556-2800

PEABODY INTERNATIONAL CORP.
PBD NYSE
4 LANDMARK SQUARE
STAMFORD CT 06904
203/348-0000

PENTA SYSTEMS INTER'L, INC.
PSLI NASDAQ
1511 GUILFORD AVENUE
BALTIMORE MD 21202
301/685-7258

PERCEPTRONICS, INCORPORATED
PERC NASDAQ
6271 VARIEL AVENUE
WOODLAND HILLS CA 91367
213/884-7470

PERFECTDATA CORPORATION
PERF NASDAQ
9174 DEERING AVENUE
CHATSWORTH CA 91311
213/998-2400

PERKIN-ELMER CORPORATION
PKN NYSE
MAIN AVENUE
NORWALK CT 06856
203/762-1000

PERSONAL COMPUTER PRODUCTS, INC
PCPI NASDAQ
16776 BERNARDO CENTER DRIVE
SAN DIEGO CA 92128
619/485-8411

PERSONAL DIAGNOSTICS, INC.
PERS NASDAQ
628 ROUTE 10
WHIPPANY NJ 07981
201/884-2034

PHASER SYSTEMS, INCORPORATED
PHAS NASDAQ
24 CALIFORNIA ST
SAN FRANCISCO CA 94111
415/434-3990

PHONE-A-GRAM SYSTEMS, INC.
PHOG NASDAQ
ONE SOUTH PARK
SAN FRANCISCO CA 94107
415/433-4170

PHOTOGRAPHIC SCIENCES CORP.
PSCX NASDAQ
P.O. BOX 338
WEBSTER NY 14580
303/591-0101

PHOTRONICS CORPORATION
PHOT NASDAQ
45 ADAMS AVENUE
HAUPPAUGE NY 11787
516/231-9500

PICO PRODUCTS, INCORPORATED
PPI ASE
103 COMMERCE BLVD
LIVERPOOL NY 13088
315/451-7700

PINETREE COMPUTER SYSTEMS, INC.
PNTR NASDAQ
8600 FREEPORT PKWAY, STE 2000
DALLAS TX 75261
214/659-9510

PIONEER HI-BRED INTERNATIONAL
PHYB NASDAQ
1206 MULBERRY STREET
DES MOINES IA 50308
515/245-3500

PIONEER-STANDARD ELECTRONICS,
PIOS NASDAQ
4800 EAST 131ST STREET
GARFIELD HEIGHTS OH 44105
216/587-3600

PITNEY-BOWES, INCORPORATED
PBI NYSE
WALTER H. WHEELER, JR. DR.
STAMFORD CT 06926
203/356-5000

PKS/COMMUNICATIONS, INCORPORATED
PKSC NASDAQ
46 QUIRK ROAD
MILFORD CT 06460
203/877-8403

PLANNING RESEARCH CORPORATION
PLN NYSE
1500 PLANNING RESEARCH DRIVE
MCLEAN VA 22102
703/556-1000

PLANTRONICS, INCORPORATED
PLX NYSE
1762 TECHNOLOGY DR, STE 225
SAN JOSE CA 95110
408/998-8388

PLASMA-THERM, INCORPORATED
PTIS NASDAQ
RT 73
KRESSON NJ 08053
609/767-6120

POLICY MANAGEMENT SYSTEMS CORP.
PMSC NASDAQ
1321 LADY STREET
COLUMBIA SC 29201
803/748-2000

POLYMER RESEARCH CORP. OF AM.
PROA NASDAQ
2186 MILL AVENUE
BROOKLYN NY 11234
212/444-4300

PORTA SYSTEMS CORPORATION
PSI ASE
6901 JERICHO TURNPIKE
SYOSSET NY 11791
516/364-9300

POWELL INDUSTRIES, INCORPORATED
POWL NASDAQ
8550 MOSLEY DRIVE
HOUSTON TX 77217
713/944-6900

POWERTEC, INCORPORATED
PWTC NASDAQ
20550 NORDHOFF STREET
CHATSWORTH CA 91311
818/882-0004

PRAB ROBOTS, INCORPORATED
PRAB NASDAQ
5944 E. KILGORE ROAD
KALAMAZOO MI 49003
616/349-8761

PREMIER INDUSTRIAL CORPORATION
PRE NYSE
4500 EUCLID AVENUE
CLEVELAND OH 44103
216/391-8300

PRENTICE-HALL, INCORPORATED
PTN ASE
SYLVAN AVENUE
ENGLEWOOD CLIFFS NJ 07632
201/592-2000

PRIAM CORPORATION
PRIA NASDAQ
20 WEST MONTAGUE EXPRESSWAY
SAN JOSE CA 95134
408/946-4600

PRIME COMPUTER, INCORPORATED
PRM NYSE
PRIME PARK
NATICK MA 01760
617/655-8000

PRINTRONIX, INCORPORATED
PTNX NASDAQ
17500 CARTWRIGHT ROAD
IRVINE CA 92713
714/549-7700

PRODIGY SYSTEMS, INCORPORATED
PDGY NASDAQ
21 MERIDIAN ROAD
EDISON NJ 08820
201/321-1717

PROGRAMMING AND SYSTEMS, INC.
PSYS NASDAQ
269 WEST 40TH STREET
NEW YORK NY 10018
212/944-9200

PROLER INTERNATIONAL CORP.
PS NYSE
P. O. BOX 286
HOUSTON TX 77001
713/675-2281

PROTOCOL COMPUTERS, INC.
PCII NASDAQ
6150 CANOGA AVENUE, STE 100
WOODLAND HILLS CA 91367
213/716-5500

PSYCH SYSTEMS, INCORPORATED
PSYC NASDAQ
600 REISTERSTOWN
BALTIMORE MD 21208
301/486-2206

PYRAMID MAGNETICS, INCORPORATED
PYRM NASDAQ
9817 VARIEL AVENUE
CHATSWORTH CA 91311
213/998-0825

QUADREX CORPORATION
QUAD NASDAQ
1700 DELL AVENUE
CAMPBELL CA 95008
408/866-4510

QUALITY MICRO SYSTEMS, INC.
QMSI NASDAQ
57 S. SCHILLINGER ROAD
MOBILE AL 36608
205/343-2767

QUALITY SYSTEMS, INCORPORATED
QSII NASDAQ
17822 E. 17TH STREET
TUSTIN CA 92680
714/731-7171

QUANTECH ELECTRONICS CORP.
QANT NASDAQ
36 OAK STREET
NORWOOD NJ 07648
201/767-1320

QUANTRONIX CORPORATION
QUAN NASDAQ
225 ENGINEERS ROAD
SMITHTOWN NY 11788
516/273-6900

QUANTUM CORPORATION
QNTM NASDAQ
1804 MCCARTHY BLVD
MILPITAS CA 95035
408/262-1100

QUANTUM DIAGNOSTICS, LTD
QDLC NASDAQ
77 ARKAY DRIVE
HAUPPAUGE NY 11788
516/231-5600

QUAZON CORPORATION
QAZN NASDAQ
3330 KELLER SPRINGS ROAD
CARROLLTON TX 75006
214/385-9200

QUEST MEDICAL, INCORPORATED
QMED NASDAQ
3312 WILEY POST ROAD
CARROLLTON TX 75006
214/387-2740

QUESTECH, INCORPORATED
QTEC NASDAQ
6858 OLD DOMINION DRIVE
MCLEAN VA 22101
703/556-8666

QUESTRONICS INCORPORATED
QSTX NASDAQ
3565 SOUTHWEST TEMPLE, #5
SALT LAKE CITY UT 84115
801/262-9923

QUINTEL CORPORATION
QNTL NASDAQ
2078 E. UNIVERSITY DR.
TEMPE AZ 85281
602/894-1981

QUOTRON SYSTEMS, INCORPORATED
QUOT NASDAQ
5454 BEETHOVEN STREET
LOS ANGELES CA 90066
213/827-4600

R.V. WEATHERFORD COMPANY
WTHR NASDAQ
6921 SAN FERNANDO ROAD
GLENDALE CA 91201
213/849-3451

RADIATION TECHNOLOGY, INC.
RTCH NASDAQ
LAKE DENMARK ROAD
ROCKAWAY NJ 07866
201/625-8400

RADIATION SYSTEMS, INCORPORATED
RADS NASDAQ
1501 MORAN ROAD
STERLING VA 22170
703/450-5680

RADIOFONE CORPORATION
RDFN NASDAQ
460 SYLVAN AVENUE
ENGLEWOOD CLIFFS NJ 07632
201/569-3200

RADIONICS, INCORPORATED
RADX NASDAQ
PO BOX 80012
SALINAS CA 93912
408/757-8877

RAGEN CORPORATION
RAGN NASDAQ
9 PORETE AVENUE
NORTH ARLINGTON NJ 07032
201/997-1000

RAMTEK CORPORATION
RMTK NASDAQ
2211 LAWSON LANE
SANTA CLARA CA 95050
408/988-2211

RANCO, INCORPORATED
RNI NYSE
P.O. BOX 248
DUBLIN OH 43017
614/764-3733

RAND CAPITAL CORPORATION
RAND NASDAQ
1300 RAND BLDG
BUFFALO NY 14203
716/853-0802

RANSBURG CORPORATION
RBG ASE
3939 W. 56TH STREET
INDIANAPOLIS IN 46254
317/298-5000

RAYCHEM CORPORATION
RYC NYSE
300 CONSTITUTION DRIVE
MENLO PARK CA 94025
415/361-3333

RAYMOND CORPORATION
RAYM NASDAQ
GREENE NY 13778
607/656-2311

RAYMOND INDUSTRIES
RAE ASE
217 SMITH STREET
MIDDLETOWN CT 06457
203/632-1800

RAYTHEON COMPANY
RTN NYSE
141 SPRING STREET
LEXINGTON MA 02173
617/862-6600

RB ROBOT CORPORATION
RBRC NASDAQ
18301 WEST 10TH AVENUE #310
GOLDEN CO 80401
303/279-5525

RCA CORPORATION
RCA NYSE
30 ROCKEFELLER PLAZA
NEW YORK NY 10020
212/621-6000

REACH, INCORPORATED
BEEP NASDAQ
301 SOUTH 68TH ST
LINCOLN NE 68510
402/483-7518

RECOGNITION EQUIPMENT, INC.
REC NYSE
2701 E. GRAUWYLER ROAD
IRVING TX 75061
214/579-6000

REFAC TECHNOLOGY DEV. CORP.
REFC NASDAQ
122 EAST 42ND STREET
NEW YORK NY 10168
212/687-4741

REFLECTONE, INCORPORATED
RFTN NASDAQ
5125 TAMPA WEST BLVD.
TAMPA FL 33614
813/885-7481

REGENCY ELECTRONICS
RGCY NASDAQ
7707 RECORDS STREET
INDIANAPOLIS IN 46226
317/545-4282

REID-ASHMAN, INCORPORATED
REAS NASDAQ
590 LAURELWOOD ROAD
SANTA CLARA CA 95050
408/727-6706

RELIABILITY, INCORPORATED
REAL NASDAQ
P.O. BOX 218370
HOUSTON TX 77218
713/492-0550

REMINGTON RAND CORPORATION
REMR NASDAQ
9950 WEST 74 STREET
MINNEAPOLIS MN 55344
612/941-0400

RENAL SYSTEMS, INCORPORATED
RENL NASDAQ
14905 28TH AVENUE NORTH
MINNEAPOLIS MN 55441
612/553-3300

RESEARCH, INCORPORATED
RESR NASDAQ
BOX 24064
MINNEAPOLIS MN 55424
612/941-3300

RESEARCH-COTTRELL, INCORPORATED
RC NYSE
P.O. BOX 1500
SOMERVILLE NJ 08876
201/685-4000

REUTER, INCORPORATED
REUT NASDAQ
410 11TH AVENUE, SOUTH
HOPKINS MN 55343
612/935-6921

REXON, INCORPORATED
REXN NASDAQ
5800 UPLANDER WAY
CULVER CITY CA 90230
213/641-7110

REYNOLDS & REYNOLDS CO. (THE)
REYNA NASDAQ
800 GERMANTOWN STREET
DAYTON OH 45407
513/443-2000

RIBI IMMUNOCHEM RESEARCH, INC.
RIBI NASDAQ
PO BOX 1409
HAMILTON MT 59840
406/363-6214

ROBINSON NUGENT, INCORPORATED
RNIC NASDAQ
P.O. BOX 470
NEW ALBANY IN 47150
812/945-0211

ROBOTIC VISION SYSTEMS, INC.
ROBV NASDAQ
536 BROADHOLLOW ROAD
MELVILLE NY 11747
516/694-8910

ROCKCOR, INCORPORATED
ROCK NASDAQ
YORK CENTER
REDMOND WA 98052
206/885-5000

ROCKWELL INTERNATIONAL CORP.
ROK NYSE
600 GRANT STREET
PITTSBURGH PA 15219
412/565-2000

ROGERS CORPORATION
ROG ASE
ONE TECHNOLOGY DRIVE
ROGERS CT 06263
203/774-9605

ROHR INDUSTRIES, INCORPORATED
RHR NYSE
FOOT OF H STREET
CHULA VISTA CA 92012
714/575-4111

ROLLINS ENVIRONMENTAL SERVICES
REN NYSE
2200 CONCORD PIKE
WILMINGTON DE 19803
302/429-2700

ROLM CORPORATION*
RM NYSE
4900 OLD IRONSIDES DRIVE
SANTA CLARA CA 95050
408/988-2900

SAFEGUARD SCIENTIFICS, INC.
SFE NYSE
630 PARK AVENUE
KING OF PRUSSIA PA 19406
215/265-4000

SAFEGUARD BUSINESS SYS. INC.
SGB NYSE
400 MARYLAND DRIVE
FORT WASHINGTON PA 19034
215/641-5000

SAGE LABORATORIES, INCORPORATED
SLAB NASDAQ
3 HURON DRIVE
NATICK MA 01760
617/653-0844

SANDERS ASSOCIATES, INC.
SAA NYSE
DANIEL WEBSTER HIGHWAY SOUTH
NASHUA NH 03061
603/885-4321

SARGENT-WELCH SCIENTIFIC CO.
SWS NYSE
7300 NORTH LINDEN AVENUE
SKOKIE IL 60077
312/677-0600

SATELCO, INCORPORATED
STEL NASDAQ
1 SATELCO PLAZA
SAN ANTONIO TX 78205
512/226-5454

SATELLINK CORPORATION
SLNK NASDAQ
1 INVERNESS DR EAST
ENGLEWOOD CO 80112
303/790-1500

SATELLITE SYNDICATED SYS., INC.
SSSN NASDAQ
8252 SOUTH HARVARD AVENUE
TULSA OK 74137
918/481-0881

SCA SERVICES, INCORPORATED*
SCV NYSE
60 STATE STREET
BOSTON MA 02109
617/367-8300

SCAN-OPTICS, INCORPORATED
SOCR NASDAQ
22 PRESTIGE PARK EAST
HARTFORD CT 06108
203/289-6001

SCAN-TRON CORPORATION
SCNN NASDAQ
3398 E. 70TH STREET
LONG BEACH CA 90805
213/633-4051

SCHAAK ELECTRONICS, INC.
SHAA NASDAQ
1415 MENDOTA HEIGHTS ROAD
ST. PAUL MN 55120
612/454-6830

SCHLUMBERGER LIMITED
SLB NYSE
277 PARK AVENUE
NEW YORK NY 10172
212/350-9400

SCHOLASTIC, INCORPORATED
SCHL NASDAQ
50 W. 44TH STREET
NEW YORK NY 10036
212/505-3000

SCI SYSTEMS, INCORPORATED
SCIS NASDAQ
5000 TECHNOLOGY DRIVE
HUNTSVILLE AL 35805
205/882-4800

SCI-MED LIFE SYSTEMS, INC.
SMLS NASDAQ
13000 COUNTY RD 6
MINNEAPOLIS MN 55441
612/559-9504

SCIENTEX CORPORATION
SCTX NASDAQ
8507 SUNSET BLVD
LOS ANGELES CA 90069
213/657-8851

SCIENTIFIC MICRO SYSTEMS
SMSI NASDAQ
777 E. MIDDLEFIELD RD.
MOUNTAIN VIEW CA 94043
415/964-5700

SCIENTIFIC COMPUTERS, INC.
SCIE NASDAQ
10101 BREN ROAD EAST
MINNETONKA MN 55343
612/933-4200

SCIENTIFIC SYS. SERVICES, INC.
SSSV NASDAQ
2000 COMMERCE DRIVE
MELBOURNE FL 32901
305/725-1300

SCIENTIFIC, INCORPORATED
SCIT NASDAQ
1703 E. SECOND STREET
SCOTCH PLAINS NJ 07076
201/322-6767

SCIENTIFIC LABORATORIES, INC.
SILA NASDAQ
P.O. BOX 5180
DENVER CO 80217
303/329-0113

SCIENTIFIC-ATLANTA, INC.
SFA NYSE
1 TECHNOLOGY PARKWAY/BOX 105600
ATLANTA GA 30348
404/441-4000

SCIENTIFIC SOFTWARE CORPORATION
SSFT NASDAQ
633 17TH STREET
DENVER CO 80202
303/571-1111

SCIENTIFIC RADIO SYSTEMS, INC.
SCRD NASDAQ
367 ORCHARD STREET
ROCHESTER NY 14606
716/458-3733

SCIENTIFIC INDUSTRIES, INC.
SCND NASDAQ
70 ORVILLE DR/AIRPORT INTL PLZ
BOHEMIA NY 11716
516/567-4700

SCIENTIFIC LEASING INC.
SG ASE
790 FARMINGTON AVENUE
FARMINGTON CT 06032
203/677-8700

SCOPE, INCORPORATED
SCPE NASDAQ
1860 MICHAEL FARADAY DRIVE
RESTON VA 22090
703/471-5600

SCOTT INSTRUMENTS CORPORATION
SCTI NASDAQ
1111 WILLOW SPRINGS DR
DENTON TX 76201
817/387-9514

SEAGATE TECHNOLOGY
SGAT NASDAQ
360 EL PUEBLO ROAD
SCOTTS VALLEY CA 95066
408/438-6550

SECURITY TAG SYSTEMS, INC.
STAG NASDAQ
P.O. BOX 23000
ST. PETERSBURG FL 33742
813/576-6399

SEEQ TECHNOLOGY, INCORPORATED
SEEQ NASDAQ
1849 FORTUNE DRIVE
SAN JOSE CA 95131
408/262-5041

SEGA ENTERPRISES, INC.
SEGA NASDAQ
2029 CENTURY PARK EAST
LOS ANGELES CA 90067
213/557-1700

SEI CORPORATION
SEIC NASDAQ
680 E. SWEDESFORD ROAD
WAYNE PA 19087
215/687-1700

SEMICON, INCORPORATED
SEME NASDAQ
10 NORTH AVENUE
BURLINGTON MA 01803
617/229-6290

SENSORMATIC ELECTRONICS CORP.
SNSR NASDAQ
500 NW 12TH AVE/HILLSBORO PLAZA
DEERFIELD BEACH FL 33341
305/427-9700

SERVAMATIC SOLAR SYSTEMS, INC.
SSSI NASDAQ
1641 CHALLENGE DR
CONCORD CA 94520
415/680-8853

SFE TECHNOLOGIES
SFEM NASDAQ
PO BOX 351
SAN FERNANDO CA 91341
213/361-1176

SFN COMPANIES, INCORPORATED
SFN NYSE
1900 EAST LAKE AVENUE
GLENVIEW IL 60025
312/998-5800

SHARED MEDICAL SYSTEMS CORP.
SMED NASDAQ
51 VALLEY STREAM PARKWAY
MALVERN PA 19355
215/296-6300

SHELDAHL, INCORPORATED
SHEL NASDAQ
NORTHFIELD MN 55057
507/663-8000

SI HANDLING SYSTEMS, INC.
SIHS NASDAQ
P.O. BOX 70
EASTON PA 18040
215/252-7321

SIERRA RESEARCH CORPORATION
SERE NASDAQ
247 CAYUGA ROAD
BUFFALO NY 14225
716/631-6200

SIGMA RESEARCH, INCORPORATED
SIGR NASDAQ
3200 GEORGE WASHINGTON WAY
RICHLAND WA 99352
509/375-0663

SIGMA-ALDRICH CORPORATION
SIAL NASDAQ
3050 SPRUCE STREET
ST. LOUIS MO 63103
314/771-5765

SIGMAFORM CORPORATION
SGMA NASDAQ
2401 WALSH AVENUE
SANTA CLARA CA 95051
408/727-6510

SILICON ELECTRO PHYSICS, INC.
SEPI NASDAQ
ONE SILICON WAY
BRADFORD PA 16701
814/362-3586

SILICON GENERAL, INCORPORATED
SILN NASDAQ
940 DETROIT AVENUE
CONCORD CA 94518
415/686-6660

SILICON VALLEY GROUP, INC.
SVGI NASDAQ
3901 BURTON DRIVE
SANTA CLARA CA 95054
408/988-0200

SILICONIX, INCORPORATED
SILI NASDAQ
2201 LAURELWOOD ROAD
SANTA CLARA CA 95054
408/988-8000

SILTEC CORPORATION
SLTC NASDAQ
3717 HAVEN AVENUE
MENLO PARK CA 94025
415/365-8600

SILVAR-LISCO
SVRL NASDAQ
1080 MARSH ROAD
MENLO PARK CA 94025
415/324-0700

SIMMONDS PRECISION PROD., INC.
SP NYSE
150 WHITE PLAINS ROAD
TARRYTOWN NY 10591
914/631-7500

SINGER COMPANY (THE)
SMF NYSE
8 STAMFORD FORUM
STAMFORD CT 06904
203/356-4200

SIPPICAN OCEAN SYSTEMS, INC.
SOSI NASDAQ
7 BARNABAS ROAD
MARION MA 02738
617/748-1160

SLOAN TECHNOLOGY CORPORATION
SLON NASDAQ
535 E. MONTECITO STREET
SANTA BARBARA CA 93103
805/963-4431

SMITHKLINE BECKMAN CORPORATION
SKB NYSE
P.O. BOX 7929
PHILADELPHIA PA 19101
215/751-4000

SOFTECH, INCORPORATED
SOFT NASDAQ
460 TOTLEN POND ROAD
WALTHAM MA 02154
617/890-6900

SOFTWARE AG SYSTEMS GROUP, INC.
SAGA NASDAQ
11800 SUNRISE VALLEY DRIVE
RESTON VA 22091
703/860-5050

SOFTWARE PUBLISHING CORP.
SPCO NASDAQ
1901 LANDINGS DRIVE
MOUNTAIN VIEW CA 94043
415/962-8910

SOLARON CORPORATION
SLRN NASDAQ
1885 W DARTMOUTH AVENUE
ENGLEWOOD CO 80110
303/762-1500

SOLID STATE SCIENTIFIC, INC.
SSI ASE
MONTGOMERYVILLE INDUSTRIAL CNTR
MONTGOMERYVILLE PA 18936
215/855-8400

SOLITRON DEVICES, INCORPORATED
SOD ASE
1177 BLUE HERON BLVD
RIVIERA BEACH FL 33404
305/848-4311

SOUTHEASTERN CAPITAL CORP.
SOE ASE
2285 PEACHTREE RD., N.E. STE 219
ATLANTA GA 30309
404/351-2557

SPACE MICROWAVE LABS., INC.
SMLI NASDAQ
1255 NORTH DUTTON AVENUE
SANTA ROSA CA 95401
707/528-8114

SPACELINK LIMITED
SPLKA NASDAQ
5275 DTC PARKWAY
ENGLEWOOD CO 80111
303/773-3053

SPARTON CORPORATION
SPA NYSE
2400 EAST GANSON STREET
JACKSON MI 49202
517/787-8600

SPECTRA-PHYSICS, INCORPORATED
SPY NYSE
3333 N. FIRST STREET
SAN JOSE CA 95134
408/946-6080

SPECTRADYNE, INCORPORATED
SPDY NASDAQ
1331 NORTH PLANO RD
RICHARDSON TX 75081
214/234-2721

SPECTRAN CORPORATION
SPTR NASDAQ
HALL ROAD
STURBRIDGE MA 01566
617/347-2261

SPECTRUM COMM. & ELECTRON. CORP
SPCT NASDAQ
62 BETHPAGE ROAD
HICKSVILLE NY 11801
516/822-9810

SPECTRUM CONTROL, INCORPORATED
SPEC NASDAQ
8061 AVONIA ROAD
FAIRVIEW PA 16415
814/455-0966

SPERRY CORPORATION
SY NYSE
1290 AVENUE OF THE AMERICAS
NEW YORK NY 10104
212/484-4444

SPEX GROUP, INCORPORATED
SPEX NASDAQ
3880 PARK AVENUE
EDISON NJ 08820
201/549-7144

SPS TECHNOLOGIES, INCORPORATED
ST NYSE
JENKINTOWN PA 19046
215/572-3000

ST. JUDE MEDICAL, INCORPORATED
STJM NASDAQ
ONE LILLEHEI PLAZA
ST PAUL MN 55117
612/483-2000

STANDARD HAVENS, INCORPORATED
SHAV NASDAQ
8800 EAST 63RD STREET
KANSAS CITY MO 64133
816/737-0400

STANDARD LOGIC, INCORPORATED
STDL NASDAQ
2215 S. STANDARD AVENUE
SANTA ANA CA 92707
714/979-4770

STANDARD MICROSYSTEMS CORP.
SMSC NASDAQ
35 MARCUS BLVD
HAUPPAUGE NY 11787
516/273-3100

STANFORD TELECOMM., INC.
STII NASDAQ
2421 MISSION COLLEGE BOULEVARD
SANTA CLARA CA 95050
408/748-1010

STAODYNAMICS, INCORPORATED
SDYN NASDAQ
P.O. BOX 1379
LONGMONT CO 80501
303/442-4728

STAR TECHNOLOGIES, INC.
STRR NASDAQ
111 S.W. FIFTH AVENUE
PORTLAND OR 97204
503/227-2052

STARTEL CORPORATION
STAS NASDAQ
2802 ALTON AVENUE
IRVINE CA 92714
714/863-9292

STIMUTECH, INCORPORATED
STIMU NASDAQ
16262 CHANDLER ROAD
EAST LANSING MI 48823
517/332-7717

STORAGE TECHNOLOGY CORPORATION
STK NYSE
2270 SOUTH 88 STREET
LOUISVILLE CO 80027
303/673-5151

STRATUS COMPUTER, INC.
STRA NASDAQ
17 STRATHMORE ROAD
NATICK MA 01760
617/460-2000

STRYKER CORPORATION
STRY NASDAQ
420 EAST ALCOTT STREET
KALAMAZOO MI 49001
616/381-3811

SUNDSTRAND CORPORATION
SNS NYSE
4751 HARRISON AVENUE
ROCKFORD IL 61101
815/226-6000

SUPERTEX, INCORPORATED
SPTX NASDAQ
1225 BORDEAUX DRIVE
SUNNYVALE CA 94089
408/744-0100

SURVIVAL TECHNOLOGY, INC.
SURV NASDAQ
7801 WOODMONT AVENUE
BETHESDA MD 20814
301/654-2303

SWITCHCO, INCORPORATED
SXCO NASDAQ
329 ALFRED AVENUE
TEANECK NJ 07666
201/837-5100

SYBRON CORPORATION
SYB NYSE
1100 MIDTOWN TOWER
ROCHESTER NY 14604
716/546-4040

SYKES DATATRONICS, INCORPORATED
SYKE NASDAQ
100 KINGS HIGHWAY
ROCHESTER NY 14617
716/266-4000

SYM-TEK SYSTEMS, INCORPORATED
SYMK NASDAQ
3912 CALLE FORTUNADA
SAN DIEGO CA 92123
617/569-6800

SYMBION, INCORPORATED
SYMB NASDAQ
825 N. 300 WEST
SALT LAKE CITY UT 84103
801/531-7022

SYMBOL TECHNOLOGIES, INC.
SMBL NASDAQ
90 PLANT AVENUE
HAUPPAUGE NY 11787
516/231-5252

SYMBOLICS, INCORPORATED
SMBX NASDAQ
11 CAMBRIDGE CENTER
CAMBRIDGE MA 02142
617/577-7500

SYNBIOTICS CORPORATION
SBIO NASDAQ
348-B RANCHEROS DRIVE
SAN MARCOS CA 92069
619/471-0710

SYNERGEN, INCORPORATED
SYGN NASDAQ
1885 33RD STREET
BOULDER CO 80301
303/442-7094

SYNTECH INTERNATIONAL, INC.
SYNE NASDAQ
10111 MILLER RD, STE 105
DALLAS TX 75238
214/340-0379

SYNTREX, INCORPORATED
STRX NASDAQ
246 INDUSTRIAL WAY WEST
EATONTOWN NJ 07724
201/542-1500

SYSCON CORPORATION
SCON NASDAQ
1054 31ST STREET, N.W.
WASHINGTON DC 20007
202/342-4000

SYSTEM INDUSTRIES, INC.
SYSM NASDAQ
1855 BARBER LANE
MILPITAS CA 95035
408/942-1212

SYSTEM INTEGRATORS, INC.
SINT NASDAQ
P.O. BOX 13626
SACRAMENTO CA 95853
916/929-9481

SYSTEMATICS, INCORPORATED
SYST NASDAQ
212 CENTER STREET
LITTLE ROCK AR 72201
501/223-5100

SYSTEMATICS GENERAL CORP.
SYSG NASDAQ
1606 OLD OX ROAD
STERLING VA 22170
703/471-2200

SYSTEMS & COMPUTER TECH. CORP.
SCTC NASDAQ
4 COUNTRY VIEW ROAD
MALVERN PA 19355
215/647-5930

SYSTEMS ASSOCIATES, INC.
SAIN NASDAQ
412 E. BOULEVARD
CHARLOTTE NC 28236
704/333-1276

SYSTEMS INTEGRATORS, INC.
SINT NASDAQ
PO BOX 13626
SACRAMENTO CA 95853
916/929-9481

SYTEK, INCORPORATED
SYTK NASDAQ
1225 CHARLESTON ROAD
MOUNTAIN VIEW CA 94043
415/966-7300

T-BAR, INCORPORATED
TBR ASE
141 DANBURY ROAD
WILTON CT 06897
203/834-8227

TAB PRODUCTS COMPANY
TBP ASE
1400 PAGE MILL ROAD
PALO ALTO CA 94304
415/852-2400

TANDEM COMPUTERS, INCORPORATED
TNDM NASDAQ
19333 VALLCO PARKWAY
CUPERTINO CA 95014
408/725-6000

TANDON CORPORATION
TCOR NASDAQ
20320 PRAIRIE STREET
CHATSWORTH CA 91311
213/993-6644

TANDY CORPORATION
TAN NYSE
1800 ONE TANDY CENTER
FORT WORTH TX 76102
817/390-3700

TDK ELECTRONICS CO, LTD
TDK NYSE
12 HARBOR PARK DRIVE
PORT WASHINGTON NY 11050
516/625-0100

TEC, INCORPORATED
TCK ASE
P.O. BOX 5646
TUCSON AZ 85703
602/792-2230

TECH-SYM CORPORATION
TSY ASE
6430 RICHMOND AVENUE, STE 460
HOUSTON TX 77057
713/785-7790

TECH/OPS, INCORPORATED
TO ASE
ONE BEACON STREET
BOSTON MA 02108
617/523-2030

TECHNALYSIS CORPORATION
TECN NASDAQ
6700 FRANCE AVENUE, S.
MINNEAPOLIS MN 55435
612/925-5900

TECHNICAL COMMUNICATIONS CORP.
TCCO NASDAQ
100 DOMINO DRIVE
CONCORD MA 01742
617/862-6035

TECHNICOM INTERNATIONAL, INC.*
TCM ASE
23 OLD KINGS HIGHWAY S.
DARIEN CT 06820
203/655-1299

TECHNODYNE, INCORPORATED
TECK NASDAQ
98 CUTTER MILL ROAD
GREAT NECK NY 11021
516/466-5100

TECHNOLOGY, INCORPORATED
TNLG NASDAQ
1115 TALBOTT TOWER
DAYTON OH 45402
513/224-9066

TECHNOLOGY FOR COMM. INTER'L
TCII NASDAQ
1625 STIERLIN ROAD
MOUNTAIN VIEW CA 94043
415/961-9180

TECHTRAN INDUSTRIES, INC.
TECH NASDAQ
200 COMMERCE DRIVE
ROCHESTER NY 14623
716/334-9640

TEKTRONIX, INCORPORATED
TEK NYSE
4900 S.W. GRIFFITH DRIVE
BEAVERTON OR 97007
503/644-0161

TELCO SYSTEMS, INCORPORATED
TELC NASDAQ
1040 MARSH ROAD, STE 100
MENLO PARK CA 94025
415/324-4300

TELE-COMMUNICATIONS, INC.
TCOMA NASDAQ
5455 SOUTH VALENTIA WAY
ENGLEWOOD CO 80111
303/771-8200

TELEBYTE TECHNOLOGY INCORPORATED
TBTIU NASDAQ
148 NEW YORK AVENUE
HALESITE NY 11743
516/423-3237

TELECOM PLUS INTER'L, INC.
TELE NASDAQ
48-40 34TH STREET
LONG ISLAND CITY NY 11101
212/392-7700

TELECOM. SPECIALISTS (TELSPEC), INC.
TLSP NASDAQ
400 OSER AVENUE
HAUPPAUGE NY 11788
716/231-0222

TELECONCEPTS CORPORATION
TLCN NASDAQ
22 CULBRO DRIVE
WEST HARTFORD CT 6110
203/525-3107

TELEDYNE, INCORPORATED
TDY NYSE
1901 AVENUE OF THE STARS
LOS ANGELES CA 90067
213/277-3311

TELEPHONE SPECIALIST, INC.
TESP NASDAQ
4944 NORTH COUNTY ROAD 18
MINNEAPOLIS MN 55428
612/533-7556

TELESCIENCES, INCORPORATED
TSC ASE
351 NEW ALBANY ROAD
MOORESTOWN NJ 08057
609/235-6227

TELEVIDEO SYSTEMS, INCORPORATED
TELV NASDAQ
1170 MORSE AVENUE
SUNNYVALE CA 94086
408/745-7760

TELEX CORPORATION (THE)
TC NYSE
6422 E. 41 STREET
TULSA OK 74135
918/627-2333

TELLABS, INCORPORATED
TLAB NASDAQ
4951 INDIANA AVENUE
LISLE IL 60532
312/969-8800

TELXON CORPORATION
TLXN NASDAQ
3330 WEST MARKET STREET
AKRON OH 44313
216/867-3700

TENNEY ENGINEERING, INC.
TNY ASE
P.O. BOX 3142
UNION NJ 07083
212/686-7870

TERA CORPORATION
TRRA NASDAQ
2150 SHATTUCK AVENUE
BERKELEY CA 94704
415/845-5200

TERADYNE, INCORPORATED
TER NYSE
183 ESSEX STREET
BOSTON MA 02111
617/482-2700

TERAK CORPORATION
TCGS NASDAQ
14151 NORTH 76TH STREET
SCOTTSDALE AZ 85260
602/998-4800

TERMIFLEX CORPORATION
TFLX NASDAQ
18 AIRPORT ROAD
NASHUA NH 03063
603/889-3883

TERMINAL DATA (TDC) CORPORATION
TERM NASDAQ
21221 OXNARD STREET
WOODLAND HILLS CA 91367
213/887-4900

TESDATA SYSTEMS CORPORATION
TDSC NASDAQ
7921 JONES BRANCH DRIVE
MC LEAN VA 22102
703/827-4000

TESTAMATIC CORPORATION
TSTC NASDAQ
5 HEMLOCK STREET
LATHAM NY 12110
518/783-6301

TEXAS INSTRUMENTS, INCORPORATED
TXN NYSE
P.O. BOX 225
DALLAS TX 75265
214/238-2011

TEXSCAN CORPORATION
TXS ASE
2446 N. SHADELAND AVENUE
INDIANAPOLIS IN 46219
317/357-8781

THERMAL ENERGY STORAGE, INC.
TESI NASDAQ
10637 ROSELLE STREET
SAN DIEGO CA 92121
619/453-1395

THERMEDICS, INCORPORATED
THMD NASDAQ
470 WILDWOOD STREET
WOBURN MA 01888
617/938-3786

THERMO ELECTRON CORP.
TMO NYSE
101 FIRST AVE
WALTHAM MA 02254
617/890-8700

THERMODYNETICS, INCORPORATED
TDYN NASDAQ
651 DAYHILL ROAD
WINDSOR CT 06095
203/683-2005

THERMWOOD CORPORATION
THWD NASDAQ
P.O. BOX 436
DALE IN 47523
812/937-4476

THOMAS & BETTS CORPORATION
TNB NYSE
920 ROUTE 202
RARITAN NJ 08869
201/685-1600

THORATEC LABORATORIES CORP.
TTEC NASDAQ
2023 EIGHTH STREET
BERKELEY CA 94710
415/841-1213

THRESHOLD TECHNOLOGY, INC.
THRS NASDAQ
1829 UNDERWOOD BLVD
DELRAN NJ 08075
609/461-9200

TIE/COMMUNICATIONS, INC.
TIE ASE
5 RESEARCH DRIVE
SHELTON CT 06484
203/929-7373

TII INDUSTRIES, INCORPORATED
TI ASE
1375 AKRON STREET
COPIAGUE NY 11726
516/789-5000

TIMBERLINE SYSTEMS, INCORPORATED
TMBS NASDAQ
7180 SOUTHWEST FIR LOOP
PORTLAND OR 97223
503/684-3660

TIME ENERGY SYSTEMS, INC.
TIME NASDAQ
2900 WILCREST DRIVE
HOUSTON TX 77042
713/780-8532

TIME SHARING RESOURCES, INC.
TIMS NASDAQ
777 NORTHERN BLVD
GREAT NECK NY 11021
516/487-0101

TIMEPLEX, INCORPORATED
TIX NYSE
ONE COMMUNICATIONS PLAZA
ROCHELLE PARK NJ 07662
201/391-1111

TIMES FIBER COMM., INC.
TFCI NASDAQ
P.O. BOX 384
WALLINGFORD CT 06492
203/265-8500

TOCOM, INCORPORATED
TOCM NASDAQ
3301 ROYALTY ROW
IRVING TX 75062
214/438-7691

TOROTEL, INCORPORATED
TTL ASE
P.O. BOX 608
RAYMORE MO 64083
816/331-8400

TOTAL SYSTEM SERVICES, INC.
TSYS NASDAQ
P.O. BOX 120
COLUMBUS GA 31902
404/571-2387

TOUCHSTONE APPLIED SCI., INC.
TASA NASDAQ
150 CLEARBROOK ROAD
ELMSFORD NY 10523
914/592-2630

TRACOR, INCORPORATED
TRR NYSE
6500 TRACOR LANE
AUSTIN TX 78721
512/926-2800

TRANSDUCER SYSTEMS, INC.
TSIC NASDAQ
1510 DELP DRIVE
KULPSVILLE PA 19443
215/256-4611

TRANSIDYNE GENERAL CORP.
TGCO NASDAQ
3850 RESEARCH PARK DRIVE
ANN ARBOR MI 48104
313/769-1900

TRANSITRON ELECTRONIC CORP.
TREL NASDAQ
100 UNICORN PARK DRIVE
WOBURN MA 01801
617/933-9640

TRANSMATION, INCORPORATED
TRNS NASDAQ
977 MT. READ BLVD
ROCHESTER NY 14606
716/254-9000

TRANSNET CORPORATION
TRNT NASDAQ
1945 ROUTE 22
UNION NJ 07083
201/688-7800

TRANSTECHNOLOGY CORPORATION
TT ASE
15233 VENTURA BLVD #400
SHERMAN OAKS CA 91403
213/990-5920

TRC COMPANIES, INCORPORATED
TRCC NASDAQ
800 CONNECTICUT BLVD
EAST HARTFORD CT 06108
203/289-8631

TRIAD SYSTEMS CORPORATION
TRSC NASDAQ
1252 ORLEANS DRIVE
SUNNYVALE CA 94086
408/734-9720

TRIANGLE MICROWAVE, INC.
TRMW NASDAQ
60 OKNER PARKWAY
LIVINGSTON NJ 07039
201/740-0100

TRILOGY LIMITED
TRILF NASDAQ
10500 RIDGEVIEW COURT
CUPERTINO CA 95014
408/973-9333

TRIO-TECH INTERNATIONAL
TRIO NASDAQ
2040 NORTH LINCOLN STREET
BURBANK CA 91504
213/846-9200

TRION, INCORPORATED
TRON NASDAQ
P.O. BOX 760
SANFORD, NC 27330
919/775-2201

TRT COMMUNICATIONS, INC.
TRTA NASDAQ
1747 PENNSYLVANIA AVENUE N.W.
WASHINGTON DC 20006
202/862-4500

TSI, INCORPORATED
TSII NASDAQ
P.O. BOX 43394
ST. PAUL MN 55164
612/483-0900

TVI ENERGY CORPORATION
TVIE NASDAQ
3012 HERZEL PLACE
BELTSVILLE MD 20705
301/595-5252

TYCO LABORATORIES, INCORPORATED
TYC NYSE
TYCO PARK
EXETER NH 03833
603/778-7331

TYLAN CORPORATION
TYLN NASDAQ
23301 SOUTH WILMINGTON AVENUE
CARSON CA 90745
213/518-3610

TYMSHARE, INCORPORATED
TYM NYSE
20705 VALLEY GREEN DRIVE
CUPERTINO CA 95014
408/446-6000

U.S. DESIGN CORPORATION
USDC NASDAQ
5100 PHILADELPHIA WAY
LANHAM MD 20706
301/577-2880

U.S. SURGICAL CORPORATION
USSC NASDAQ
150 GLOVER AVENUE
NORWALK CT 06850
203/866-5050

U.S. TELEPHONE, INCORPORATED
USTL NASDAQ
108 S. AKARD STREET
DALLAS TX 75202
214/741-1957

ULTIMATE CORPORATION (THE)
ULT ASE
77 BRANT AVENUE
CLARK NJ 07066
201/388-8800

ULTRASYSTEMS, INCORPORATED
ULTR NASDAQ
2400 MICHELSON DRIVE
IRVINE CA 92715
714/752-7500

UNGERMANN-BASS, INC.
UNGR NASDAQ
2560 MISSION COLLEGE BLVD
SANTA CLARA CA 95050
408/496-0111

UNITED EDUCATION & SOFTWARE
UESS NASDAQ
15720 VENTURA BLVD., STE 512
ENCINO CA 91436
213/907-6649

UNITED INDUSTRIAL CORPORATION
UIC NYSE
18 E. 48TH STREET
NEW YORK NY 10017
212/752-8787

UNITED TECHNOLOGIES CORPORATION
UTX NYSE
UNITED TECHNOLOGIES BLDG
HARTFORD CT 06101
203/728-7000

UNITED TELECOMMUNICATIONS, INC.
UT NYSE
P.O. BOX 11315
KANSAS CITY MO 64112
913/676-3000

UNITED TELECONTROL ELECTRONICS
UTEL NASDAQ
3500 SUNSET AVENUE
ASHBURY PARK NJ 07712
201/922-1000

UNITRODE CORPORATION
UTR NYSE
5 FORBES ROAD
LEXINGTON MA 02173
617/861-6540

UNIVERSAL VOLTRONICS CORP.
UVOL NASDAQ
27 RADIO CIRCLE DR
MT KISCO NY 10549
914/241-1300

UNIVERSAL SECURITY INSTR., INC.
USEC NASDAQ
10324 S. DOLFIELD ROAD
OWINGS MILLS MD 21117
301/363-3000

UNIVERSITY GENETICS COMPANY
UGEN NASDAQ
537 NEWTOWN AVENUE
NORWALK CT 06852
203/856-9012

URI THERM-X, INCORPORATED
URTXU NASDAQ
375 N. BROADWAY
JERICHO NY 11753
516/681-7880

UTL CORPORATION
UTLC NASDAQ
4500 WEST MOCKINGBIRD LANE
DALLAS TX 75209
214/350-7601

V BAND SYSTEMS, INCORPORATED
VBAN NASDAQ
5 O'DELL PLAZA
YONKERS NY 10701
914/964-0900

VALID LOGIC SYSTEMS, INC.
VLID NASDAQ
2820 ORCHARD PARKWAY
SAN JOSE CA 95134
408/945-9400

VARIAN ASSOCIATES, INCORPORATED
VAR NYSE
611 HANSEN WAY
PALO ALTO CA 94303
415/493-4000

VARO, INCORPORATED
VRO NYSE
P.O. BOX 401426
GARLAND TX 75040
214/272-1571

VECTOR AUTOMATION, INCORPORATED
VCTA NASDAQ
VILLAGE OF CROSS KEYS
BALTIMORE MD 21210
301/433-4200

VECTOR GENERAL, INCORPORATED
VGEN NASDAQ
21300 OXNARD STREET
WOODLAND HILLS CA 91367
213/346-3410

VECTOR GRAPHIC, INCORPORATED
VCTR NASDAQ
500 N. VENTIC PARK ROAD
THOUSAND OAKS CA 91320
805/499-5831

VEECO INSTRUMENTS, INCORPORATED
VEE NYSE
515 BROAD HOLLOW ROAD
MELVILLE NY 11747
516/694-4200

VEGA BIOTECHNOLOGIES, INC.
VEGA NASDAQ
1250 E. AERO PARK BLVD
TUCSON AZ 85706
602/746-1401

VERBATIM CORPORATION
VRB ASE
323 SOQUEL WAY
SUNNYVALE CA 94086
408/245-4400

VERDIX CORPORATION
VRDX NASDAQ
7655 OLD SPRINGHOUSE ROAD
MCLEAN VA 22102
703/448-1980

VERMONT RESEARCH CORPORATION
VRE ASE
PRECISION PARK
N. SPRINGFIELD VT 05150
802/886-2256

VERNITRON CORPORATION
VRN ASE
2001 MARCUS AVE
LAKE SUCCESS NY 11042
516/775-8200

VICON FIBEROPTICS CORPORATION
VFOX NASDAQ
90 SECOR LANE
PELHAM MANOR NY 10803
914/738-5006

VICON INDUSTRIES, INCORPORATED
VII ASE
125 E BETHPAGE ROAD
PAINVIEW NY 11803
516/293-2200

VICTOR TECHNOLOGIES, INC.
VICR NASDAQ
380 EL PUEBLO ROAD
SCOTTS VALLEY CA 95066
408/438-6680

VIDEO CORPORATION OF AMERICA
VICA NASDAQ
231 EAST 55TH STREET
NEW YORK NY 10022
212/355-1600

VIDEOVISION, INCORPORATED
VVIS NASDAQ
1116 EDGEWATER AVENUE
RIDGEFIELD NJ 07657
201/943-7860

VIRGINIA PANEL CORPORATION
VPCO NASDAQ
1400 NEW HOPE ROAD
WAYNESBORO VA 22980
703/949-8376

VISHAY INTERTECHNOLOGY
VSH ASE
63 LINCOLN HIGHWAY
MALVERN PA 19355
215/644-1300

VISUAL TECHNOLOGY, INCORPORATED
VSAL NASDAQ
540 MAIN STREET
TEWKSBURY MA 01876
617/851-5000

VIVIGEN INCORPORATED
VIVI NASDAQ
550 ST MICHAELS DRIVE
SANTA FE NM 87501
505/988-9744

VLSI TECHNOLOGY, INC.
VLSI NASDAQ
1101 MCKAY PLACE
SAN JOSE CA 95131
408/942-1810

VM SOFTWARE, INCORPORATED
VMSI NASDAQ
2070 CHAIN BRIDGE RD
VIENNA VA
703/821-6886

VMX, INCORPORATED
VMXI NASDAQ
1241 COLUMBIA DRIVE
RICHARDSON TX 75081
214/699-1461

VOICEMAIL INTERNATIONAL, INC.
VOIC NASDAQ
3350 SCOTT BLVD. # 49
SANTA CLARA CA 95051
408/496-6555

VOLT INFO. SCIENCES INC.
VOLT NASDAQ
1221 AVENUE OF THE AMERICAS
NEW YORK NY 10020
212/309-0200

WABASH DATATECH, INCORPORATED
DATK NASDAQ
1701 GOLF ROAD, SUITE 400
ROLLING MEADOWS IL 60008
312/593-6363

WALLACE COMPUTER SERVICES, INC.
WCS NYSE
4600 W. ROOSEVELT BLVD
HILLSIDE IL 60162
312/626-2000

WANG LABORATORIES, INCORPORATED
WAN.B ASE
ONE INDUSTRIAL AVENUE
LOWELL MA 01851
617/459-5000

WARNER COMMUNICATIONS, INC.
WCI NYSE
75 ROCKEFELLER PLAZA
NEW YORK NY 10019
212/484-8000

WASTE MANAGEMENT, INCORPORATED
WMX NYSE
3003 BUTTERFIELD ROAD
OAK BROOK IL 60521
312/654-8800

WATERS INSTRUMENTS, INC.
WTRS NASDAQ
P.O. BOX 6117
ROCHESTER MN 55903
507/288-7777

WATKINS-JOHNSON COMPANY
WJ NYSE
333 HILLVIEW ROAD
PALO ALTO CA 94304
415/493-4141

WAVETEK CORPORATION
WVTK NASDAQ
P.O. BOX 85265
SAN DIEGO CA 92138
619/279-2200

WELLS-GARDNER ELECTRONICS CORP.
WGA ASE
2701 N. KILDARE AVENUE
CHICAGO IL 60639
312/252-8220

WESPERCORP
WP ASE
14321 NEW MYFORD ROAD
TUSTIN CA 92680
714/730-6250

WESTERN DIGITAL CORPORATION
WDCL NASDAQ
2445 MCCABE WAY
IRVING CA 92714
714/557-3550

WESTERN MICROWAVE, INC.
WMIC NASDAQ
1271 REAMWOOD AVENUE
SUNNYBALE CA 94089
408/734-1631

WESTERN MICRO TECHNOLOGY
WSTM NASDAQ
10040 BUBB ROAD
CUPERTINO CA 95014
408/725-1660

WESTINGHOUSE ELECTRIC CORP.
WX NYSE
GATEWAY CENTER
PITTSBURG PA 15222
412/255-3800

WHITTAKER CORPORATION
WKR NYSE
10880 WILSHIRE BLVD
LOS ANGELES CA 90024
213/475-9411

WICAT SYSTEMS, INC.
WCAT NASDAQ
1875 S. STATE STREET
OREM UT 84057
801/224-6400

WIDERGREN COMMUNICATIONS, INC.
WIDE NASDAQ
1500 EAST HAMILTON
CAMPBELL CA 95008
408/377-9981

WILEY (JOHN) AND SONS, INC.
WILLB NASDAQ
605 THIRD AVENUE
NEW YORK NY 10016
212/850-6000

WILFRED AMERICAN ED. CORP.
WAE NYSE
1657 BROADWAY
NEW YORK NY 10019
212/582-6690

WILLIAMS ELECTRONICS, INC.
WMS NYSE
3401 NORTH CAROLINA AVENUE
CHICAGO IL 60618
312/267-2240

WIND BARON CORPORATION
WNBR NASDAQ
3702 WEST LOWER BUCKEYE RD
PHOENIX AZ 85009
602/269-6900

WORLD OF COMPUTERS, INC.
WOCO NASDAQ
6600 EAST HAMPTON AVE STE 200
DENVER CO 80224
303/759-5555

WYLE LABORATORIES
WYL NYSE
128 MARYLAND STREET
EL SEGUNDO CA 90245
213/678-4251

WYLY CORPORATION
WLY NYSE
UCC TOWER, EXCHANGE PARK
DALLAS TX 75235
214/353-7100

WYSE TECHNOLOGY
WYSE NASDAQ
3040 N. FIRST ST.
SAN JOSE CA 95134
408/946-3075

XEBEC
XEBC NASDAQ
432 LAKESIDE DRIVE
SUNNYVALE CA 94086
408/733-4200

XEROX CORPORATION
XRX NYSE
P.O. BOX 1600
STAMFORD CT 06904
203/329-8700

XICOR, INCORPORATED
XICO NASDAQ
851 BUCKEYE COURT
MILPITAS CA 95035
408/946-6920

XIDEX CORPORATION
XIDX NASDAQ
2141 LANDINGS DRIVE
MOUNTAIN VIEW CA 94043
415/965-7350

XL/DATACOMP, INCORPORATED
XLDC NASDAQ
907 N. ELM STREET
HINSDALE IL 60521
312/323-1200

ZENITH RADIO CORPORATION
ZE NYSE
1000 MILWAUKEE AVENUE
GLENVIEW IL 60025
312/391-7000

ZENTEC CORPORATION
ZENT NASDAQ
2400 WALSH AVENUE
SANTA CLARA CA 95050
408/727-7662

ZETA LABORATORIES, INC.
ZETA NASDAQ
3265 SCOTT BLVD
SANTA CLARA CA 95051
408/727-6001

ZEUS COMPONENTS, INC.
ZUSC NASDAQ
100 MIDLAND AVENUE
PORT CHESTER NY 10573
914-937-7400

ZITEL CORPORATION
ZITL NASDAQ
235 CHARCOT AVENUE
SAN JOSE CA 95131
408/946-9600

ZIYAD, INCORPORATED
ZYAD NASDAQ
100 FORD ROAD
DENVILLE NJ 07834
201/627-7600

ZONIC CORPORATION
ZNIC NASDAQ
2000 FORD CIRCLE
MILFORD OH 45150
513/248-1911

ZURN INDUSTRIES
ZRN NYSE
1 ZURN PLACE, BOX 2000
ERIE PA 16512
814/452-2111

ZYCAD CORPORATION
ZCAD NASDAQ
1315 RED FOX ROAD
ARDEN HILLS MN 55112
612/631-3175

ZYMOS CORPORATION
ZMOS NASDAQ
477 NORTH MATHILDA AVENUE
SUNNYVALE CA 94068
408/730-8800

ZYTREX CORPORATION
ZTRX NASDAQ
750 EAST ARQUES AVENUE
SUNNYVALE CA 94086
408/733-3973

SELECTED FOREIGN HIGH TECH COMPANIES

ADVANCED SEMICONDUCTOR NV
ASMIF NASDAQ
4302 EAST BROADWAY
PHOENIX AZ 85040
602/243-4221

AJINOMOTO CO., INC.
TOKYO
1-5-8, KYOBASHI
CHUOKU TOKYO 104 JAPAN

ASEA, AB
ASEAY NASDAQ
P.O. BOX 7373
S-103 91 STOCKHOLM SWEDEN
+46/824-5950

BAXTER TECHNOLOGIES CORPORATION
BTC MONT
69 MONTCALM AVENUE
TORONTO ONT CANADA

BIOGEN N.V.
BGENF NASDAQ
FOURTEEN CAMBRIDGE CENTER
CAMBRIDGE MA 02142
617/864-8900

BRITISH TELECOMMUNICATIONS
BTY NYSE
2-12 GRESHAM STREET
LONDON ENGLAND EC2V 7AG
441/357-3000

CABLESHARE, INCORPORATED
CSH TORONTO
20 ENTERPRISE DRIVE
LONDON ONT CANADA N6A4L6
519/686-2900

CAE INDUSTRIES, LIMITED
CAE MONT
P.O. BOX 30, ROYAL BANK PLAZA
TORONTO ONT CANADA M5J 2J1

CANADIAN MARCONI COMPANY
CMC TORONTO
2442 TRENTON AVENUE
MONTREAL QUE CANADA H3P 1Y9
514/341-7630

DATACROWN, INCORPORATED
OTC
650 MCNICOLL AVENUE
WILLOWDALE ONTARIO CANADA M2H 2E1
416/499-1012

DATALINE, INCORPORATED
DLS TORONTO
175 BEDFORD ROAD
TORONTO ONTARIO CANADA M5R 2L2
416/964-9515

DATATECH SYSTEMS LIMITED
DTK VANC
1095 MCKENZIE AVENUE
VICTORIA B.C. CANADA V8P 2L5
604/479-7117

DEVELCON ELECTRONICS LTD
DLC TORONTO
856-51ST STREET E.
SASKATOON SASK CANADA S7K5C7
306/664-3777

ELBIT COMPUTERS LTD.
NASDAQ
P.O. BOX 5390
HAIFA 31053 ISRAEL
(04)524222

ELECTRONICS CORP OF ISRAEL
ECILF NASDAQ
88 GIBOREI ISRAEL ST.
TEL AVIV 67891 ISRAEL
972/333-3241

ELRON ELECTRONIC IND., INC.
ELRNZ NASDAQ
1120 CONNECTICUT AVE, STE 1124
WASHINGTON DC 20036

ELSCINT, LIMITED
ELSTF NASDAQ
ADVANCED TECH' CNTR' P.O.B. 5258
HAIFA 31052 ISRAEL
212/838-3341

ENS BIO LOGICALS, INCORPORATED
BIOLF NASDAQ
7 HINTON AVENUE, NORTH
OTTAWA, ONT CANADA K1Y 4P1
416/686-4328

EPITEK INTERNATIONAL, INC.
EPK TORONTO
100 SCHNEIDER ROAD
KANATA ONTARIO CANADA K2K 1Y2
613/592-2240

ERICSSON (LM) TELEPHONE CO.
ERICY NASDAQ
S-126 25
STOCKHOLM SWEDEN

FANUC LIMITED
TOKYO
1331 GREENLEAF AVENUE
ELK GROVE VILLAGE IL 60007
312/364-5060

FUJITSU LIMITED
NASDAQ
680 FIFTH AVENUE
NEW YORK NY 10019
212/265-5360

G & B AUTOMATED EQUIPMENT LTD.
GB TORONTO
580 SUPERTEST ROAD
DOWNSVIEW ONT CANADA M3J 2M7
416/661-1500

GANDALF TECHNOLOGIES, INC.
GANDF NASDAQ
9 SLACK ROAD
NEPEAN ONT CANADA K2G OB7
613/225-0565

GEAC COMPUTER CORPORATION LTD
GAC TORONTO
350 STEELCASE ROAD WEST
MARKHAM ONT CANADA L3R 1B3
416/475-0525

GLENAYRE ELECTRONICS LIMITED
GLNT TORONTO
1551 COLUMBIA STREET
NORTH VANCOUVER B.C. CANADA V7J 1A3
604/980-6041

HELIX CIRCUITS, INCORPORATED
HLX TORONTO
8421 DARNLEY ROAD
MONTREAL QUEBEC CANADA

HITACHI LTD
HIT NYSE
437 MADISON AVENUE
NEW YORK NY 10022
212/758-5420

INSTRUMENTARIUM
INMRY NASDAQ
ELIMAENKATU 22-24, SF-00510
HELSINKI 51 FINLAND
711211

INTER'L PHASOR TELECOM LTD.
IPTFC NASDAQ
134 ABBOTT STREET
VANCOUVER BC CANADA V6B2K6
604/683-7636

INTERPHARM LABORATORIES LIMITED
IPLLF NASDAQ
SCIENCE IND. PARK/KIRYAT WEIZMAN
NESZIONA ISRAEL

KYOWA HAKKO KOGYO CO., LTD.
TOKYO
6-1 OHTEMACHI 1-CHOME
CHIYODA-KU TOKYO 100 JAPAN

LANPAR TECHNOLOGIES, INC.
LPR MONT
85 TORBAY ROAD
MARKHAM ONTARIO CANADA L3R 1G7
416/475-9123

LASER INDUSTRIES LIMITED
LAS ASE
NEVE SHARETT
TEL AVIV 61131 ISRAEL
212/697-8840

LASER-SCAN INTERNATIONAL, LTD
LASE NASDAQ
CAMBRIDGE SCIENCE PARK
MILTON ROAD CAMBRIDGE ENGLAND CB4 4BH
(0223)-315414

LEIGH INSTRUMENTS LIMITED
LHI TORONTO
2680 QUEENS VIEW DRIVE
OTTAWA, ONTARIO CANADA K2B 8J9
613/820-9720

LINEAR TECHNOLOGY, INCORPORATED
LTI TORONTO
970 FRASER DRIVE
BURLINGTON ONTARIO CANADA L7R 3Y3
416/632-2996

LUMONICS, INCORPORATED
LUM TORONTO
105 SCHNEIDER ROAD
KANATA ONT CANADA K2K 1Y3
613/592-1490

MATSUSHITA ELECTRIC CO. LTD
MC NYSE
KADOMA
OSAKA JAPAN
201/348-7705

MEMOTEC DATA, INCORPORATED
OTC
4940 FISHER STREET
MONTREAL QUEBEC CANADA H4T 1J6
514/738-4781

MITEL CORPORATION
MLT NYSE
350 LEGGETT DRIVE
KANATA ONT CANADA K2K 1X3
613/592-2122

NABU MANUFACTURING CORPORATION
NBK MONT
1051 BAXTER ROAD
OTTAWA ONTARIO CANADA K2C 3P2
613/596-6700

NATIONAL BUSINESS SYSTEMS
NBSIF NASDAQ
3220 ORLANDO DRIVE
MISSISSAUGA ONTARIO CANADA L4V 1R5
416/671-3334

NEC CORPORATION
NIPNY NASDAQ
33-1 SHIBA 5-CHOME
MINATO-KU TOKYO 108 JAPAN

NELMA INFORMATION, INC.
OTC
5170A TIMBERLEA BLVD
MISSISSAUGA ONTARIO CANADA L4W 2S5
416/624-0334

NORSAT INTERNATIONAL, INC.
NIN VANC
302 - 12886 78TH AVENUE
SURREY B.C. CANADA V3W 8E7
604/591-3334

NORSK DATA A.S.
NORKZ NASDAQ
55 WILLIAM STREET
WELLESLEY MA 02181
617/237-7945

NORTHERN TELECOM LIMITED
NT NYSE
33 CITY CENTRE DRIVE
MISSISSAUGA ONT CANADA L5B 2N
416/275-0960

NOVO INDUSTRI A/S
NVO NYSE
485 MADISON AVENUE
NEW YORK NY 10022
212/371-2200

OCEANIC ELECTRONICS CORPORATION
OTC
1155 SHERBROOKE ST W. STE 1601
MONTREAL QUEBEC CANADA H3A 2N3
514/288-6422

OLIVETTI & C., (ING. C.) SpA
NASDAQ
VIA JERVIS 77
IVREA ITALY

OMNIBUS COMPUTER GRAPHICS, INC.
OMI TORONTO
2180 YONGE STREET
TORONTO ONT CANADA M4T 2T1
416/489-6020

OPTROTECH LTD.
OPTRF NASDAQ
INDUSTRIAL ZONE B, BOX 69
NES ZIONA 70450 ISRAEL
(054) 71896

ORCATECH, INCORPORATED
ORC TORONTO
2680 QUEENSVIEW DRIVE
OTTAWA ONT CANADA K2B 8H6
613/820-9602

PATRICK COMPUTER SYSTEMS, INC.
PCS VANC
11 PLYMOUTH ST
WINNIPEG MANITOBA CANADA R2X 2V5
204/632-9128

PHARMACIA AB
PHABY NASDAQ
BOX 604
S-75125 UPPSALA SWEDEN
+46 18 163000

PHILIPS NV
PGLOY NASDAQ
GROENEWOUDSEWEG 1
5621 BA EINDHOVEN THE NETHERLANDS
314/073-2312

PLESSEY COMPANY PLC (THE)
PLY NYSE
2-60 VICARAGE LANE
ILFORD ESSEX ENGLAND I614AQ
212/752-4441

PRIMROSE TECHNOLOGY CORP.
PTE VSE
P.O. BOX 10368
VANCOUVER BC CANADA V7Y 1G5
604/682-2296

RACAL ELECTRONICS PLC.
NASDAQ
WESTERN ROAD
BRACKNELL BERKSHIRE ENGLAND RG12 1RG
(0344)54119

REAL TIME DATAPRO LIMITED
OTC
797 DON MILLS ROAD
DON MILLS ONTARIO CANADA M3C 1V1
416/429-0440

RODIME PLC
RODMY NASDAQ
59 NASMYTH RD,SOUTHFIELD IND.EST
GLENROTHES SCOTLAND KY62SD

ROGERS CABLESYSTEMS, INC.
RCI.A TORONTO
COMMERCIAL UNION TOWER
TORONTO ONT CANADA M5K 1J5
416/864-2373

RONYX CORPORATION LIMITED
RXC TORONTO
P.O. BOX 125
FORT ERIE ONTARIO CANADA L2A 5M6
416/871-2100

SANYO ELECTRIC CO. LTD
SANYY NASDAQ
200 RISER ROAD
LITTLE FERRY NJ 07643
201/641-2333

SCINTREX LIMITED
SCT TORONTO
222 SNIDERCROFT ROAD
CONCORD ONT CANADA L4K 1B5
416/669-2280

SCITEX CORPORATION LIMITED
SCIXF NASDAQ
P.O. BOX 330
46103 HERZLIA B ISRAEL

SIEMENS AG
NASDAQ
WITTELSBACHERPLATZ 2
D-8000 MUNICH 2 F.R. GERMANY

SILTRONICS LIMITED
SLX TORONTO
436 HAZELDEAN ROAD
KANATA ONT CANADA K2L1T9
613/836-5003

SONY CORPORATION
SNE NYSE
9 W. 57TH STREET
NEW YORK NY 10019
212/371-5800

SPAR AEROSPACE LIMITED
SPZ TORONTO
ROYAL BANK PLAZA BOX 83
TORONTO ONT CANADA M5J 2J2
416/885-0480

SYDNEY DEVELOPMENT CORPORATION
SYN TORONTO
1385 W. 8TH AVENUE
VANCOUVER B.C. CANADA V6H 3B9
604/734-8822

SYSTEMHOUSE
SHS.A MONT
90 SPARKS STREET
OTTAWA ONT CANADA K1P5B4
613/236-9734

TAKEDA RIKEN CO., LIMITED
TOKYO
385 SYLVAN AVENUE
ENGLEWOOD CLIFFS NJ 07632
201/569-4114

TALOS INDUSTRIES, INCORPORATED
TLO VANC
415-625 HOWE STREET
VANCOUVER B. C. CANADA V6C 2T6
604/682-6677

TELESAT CANADA
OTC
33 RIVER ROAD
OTTAWA ONTARIO CANADA K1L 8B9
613/746-5920

TELESCAN ELECTRONICS & COMM.
OTC
500, 67 RICHMOND ST WEST
TORONTO ONTARIO CANADA M5H 1Z5

TRILLIUM TELEPHONE SYS., INC.
TLM MONT
P.O. BOX 13030
KANATA ONT CANADA K2K1X3
613/592-2550

SELECTED HIGH TECH UNDERWRITERS

ALLEN & CO., INC.
711 FIFTH AVENUE
NEW YORK NY 10022
212/832-8000

BATEMAN EICHLER, HILL RICHARDS, INC.
700 S. FLOWER STREET
LOS ANGELES CA 90017
213/625-3545

BEAR, STEARNS & CO.
55 WATER STREET
NEW YORK NY 10041
212/952-5000

BECKER (A.G.) PARIBAS, INC.
55 WATER STREET
NEW YORK NY 10041
212/747-4000

BLAIR (D.H.) & CO., INC.
44 WALL STREET
NEW YORK NY 10005
212/747-0066

BLYTH EASTMAN PAINE WEBBER, INC.
1221 AVE. OF THE AMERICAS
NEW YORK NY 10020
212/730-8500

BROWN (ALEX.) & SONS
135 E. BALTIMORE ST.
BALTIMORE MD 21202
301/727-1700

BUTCHER & SINGER
211 S. BROAD ST.
PHILADELPHIA PA 19105
215/985-5000

CITIWIDE SECURITIES CORP.
111 BROADWAY
NEW YORK NY 10006
212/608-4115

DAIN BOSWORTH, INC.
100 DAIN TOWER
MINNEAPOLIS MN 55402
612/371-2711

DEAN WITTER REYNOLDS, INC.
101 CALIFORNIA STREET
SAN FRANCISCO CA 94120
415/955-6000

DONALDSON, LUFKIN & JENRETTE SEC. CORP.
140 BROADWAY
NEW YORK NY 10005
212/902-2000

DREXEL BURNHAM LAMBERT, INC.
60 BROAD STREET
NEW YORK NY 10004
212/480-6000

DUANE (JAMES) & CO., INC.
11 BROADWAY
NEW YORK NY 10004
212/248-6960

EBERSTADT (F.) & CO.
61 BROADWAY
NEW YORK NY 10006
212/480-0800

EDWARDS (A.G.) & SONS, INC.
ONE NORTH JEFFERSON ST.
ST. LOUIS MO 63103
314/289-3000

FIRST BOSTON CORP.
PARK AVENUE PLAZA
NEW YORK NY 10055
212/909-2000

FIRST WILSHIRE SECURITIES
ONE WILSHIRE BLVD.
LOS ANGELES CA 90017
213/485-8904

GILFORD SECURITIES, INC.
509 MADISON AVENUE
NEW YORK NY 10022
212/888-6400

GOLDMAN, SACHS & CO.
55 BROAD STREET
NEW YORK NY 10004
212/902-1000

HAMBRECHT & QUIST, INC.
235 MONTGOMERY STREET
SAN FRANCISCO CA 94104
415/986-5500

HUBERMAN MARGARTTEN & STRAUS, INC.
17820 W. DIXIE HIGHWAY
MIAMI BEACH FL 33160
305/935-3070

HUTTON (E.F.) & CO., INC.
ONE BATTERY PARK PLAZA
NEW YORK NY 10004
212/742-5000

KIDDER, PEABODY & CO., INC.
10 HANOVER SQUARE
NEW YORK NY 10005
212/747-2000

LADENBURG, THALMANN & CO., INC.
540 MADISON AVENUE
NEW YORK NY 10022
212/940-0100

LAIDLAW ANSBACHER, INC.
40 RECTOR STREET
NEW YORK NY 10006
212/306-6100

LAZARD FRERES & CO.
ONE ROCKEFELLER PLAZA
NEW YORK NY 10020
212/489-6600

LEHMAN BROTHERS KUHN LOEB, INC.
55 WATER STREET
NEW YORK NY 10041
212/558-1500

MERRILL LYNCH CAPITAL MARKETS
ONE LIBERTY PLAZA
NEW YORK NY 10080
212/637-7455

MONTGOMERY SECURITIES
235 MONTGOMERY STREET
SAN FRANCISCO CA 94104
415/989-2050

MORGAN STANLEY & CO., INC.
1251 AVE. OF THE AMERICAS
NEW YORK NY 10020
212/974-4000

MOSLEY, HALLGARTEN, ESTABROOK & WEEDEN
ONE NEW YORK PLAZA
NEW YORK NY 10004
212/363-6900

MULLER & CO., INC.
111 BROADWAY
NEW YORK NY 10006
212/766-1700

OPPENHEIMER & CO.
ONE NEW YORK PLAZA
NEW YORK NY 10004
212/825-4000

PAINE WEBBER
140 BROADWAY
NEW YORK NY 10004
212/437-2121

PAULSON INVESTMENT CO., INC.
729 S.W. ALDER STREET
PORTLAND OR 97205
503/243-6000

PIPER, JAFFRAY & HOPWOOD
733 MARQUETTE
MINNEAPOLIS MN 55402
612/371-6111

PRESCOTT, BALL & TURBEN, INC.
1331 EUCLID AVENUE
CLEVELAND OH 44115
216/574-7300

PRUDENTIAL-BACHE SECURITIES, INC.
100 GOLD STREET
NEW YORK NY 10038
212/791-1000

ROBERTSON, COLMAN & STEPHENS
100 CALIFORNIA STREET
SAN FRANCISCO CA 94111
415/781-9700

ROONEY, PACE, INC.
11 BROADWAY
NEW YORK NY 10004
212/908-7700

ROTAN MOSLE, INC.
1500 S. TOWER
HOUSTON TX 77002
713/236-3000

ROTHSCHILD (L.F.), UNTERBERG, TOWBIN
55 WATER STREET
NEW YORK NY 10041
212/425-3300

SALOMON BROTHERS, INC.
ONE NEW YORK PLAZA
NEW YORK NY 10004
212/747-7450

SEIDLER AMDEC SECURITIES
515 S. FIGUEROA STREET
LOS ANGELES CA 90071
213/680-0111

SHEARSON LEHMAN/AMERICAN EXPRESS
TWO WORLD TRADE CENTER
NEW YORK NY 10048
212/321-6000

SHERWOOD SECURITIES CORP.
30 MONTGOMERY STREET
JERSEY CITY NJ 07302
212/332-8881

SMITH BARNEY, HARRIS UPHAM & CO., INC.
1345 AVE. OF THE AMERICAS
NEW YORK NY 10105
212/399-6000

SUTRO & CO., INC.
201 CALIFORNIA ST.
SAN FRANCISCO CA 94111
415/445-8500

THOMSON MCKINNON SECURITIES, INC.
ONE NEW YORK PLAZA
NEW YORK NY 10004
212/482-7000

WALFORD, DEMARET & CO.
1512 LARIMER ST. SUITE 300
DENVER CO 80202
303/629-7800

WEDBUSH, NOBLE, COOKE, INC.
615 S. FLOWER STREET
LOS ANGELES CA 90030
213/620-1750

WOODMAN KIRKPATRICK & GILBREATH
ONE POST STREET
SAN FRANCISCO CA 94104
415/781-2500